An Aircraft Mechanic

When Embry Riddle Educated Mechanics
In Miami

GC Tracy

An Aircraft Mechanic

When Embry Riddle Educated Mechanics
In Miami

Chapter 1

Love at First Roar

Love at first roar for aircraft came one fine, crisp morning on a dock beside the St. Mary's River at Sault Ste. Marie, Ontario. It's what happens to the vast majority of us who continue on to aviation careers. Something ignites that first love. It could be a first flight, perhaps on an airliner, or an image in a magazine or a movie, or it could be a very up-close, deafening, shuddering, water spray in the face encounter with a roaring Norseman float plane as happened with me on that dock. I was instantly smitten.

That summer when I was seventeen, my father informed me that a flying club had been started in the Sault (pronounced Soo). It was called The Algoma Flying Club, and only operated float planes, a J3 Cub and a Taylorcraft, on the river. My father told me he was signing up for flying lessons and asked me if I wanted lessons too.

Even though I had not yet been aviation smitten at that point, I agreed to that interesting adventure. Little did I know then I would eventually be involved in Canada's transition from the manufacturing of brilliantly designed but relatively simple bush planes to equally brilliantly designed but vastly more technically sophisticated turbine airliners. And little also did I then know the real reason why I was being signed up for flying lessons. It had a lot to do with my second flight.

As for my first flight, it had been aboard a DC-3 the year previously. The reason love for airplanes had not occurred on that first flight occasion is that the necessary hormones for infatuation had been temporarily cancelled by my projectile

3

barfing the moment the wheels touched the runway on landing.

I honestly didn't mean to turn my head toward my older brother at that moment. But he, my brother, was wise in the sordid ways of this world. He had somehow perceived what was coming. As an act of self-preservation, he had already extracted a barf bag from the pouch of the seat in front of him. At the critical moment, he suddenly yanked it open to safely receive my vile gift as if he was a baseball catcher.

The second flight, the one that led to love at first roar, occurred as I drove with a friend in my father's brand new, gleaming MGA sports car along a highway north of the Sault with the convertible top down. I must emphasize that I was exactly at the posted speed limit. Honest. Then, in the middle of a turn in the road, I was shocked to see another car rushing directly at me in my lane. I guess the other driver was either drunk or distracted. He was never caught for his wicked deed.

I aimed for the shoulder of the road, and that's all I remember. Apparently the road's shoulder was soft. It gripped the right front tire. The fellow in the car following behind me later described to me what happened. He said that as the MG flipped, my friend fell out for a soft, safe landing in the tall grass. There were no seat belts in those days except in race cars and airplanes. The steering wheel held me in. When the car made a second, more violent flip, I was catapulted into the air. The witness added that my body sailed very gracefully on that, my second flight. I disappeared from his view into the trees and bushes.

It took a while for me to be found. Amazingly, I missed hitting any trees. Waking after a bit, I discovered that I had been fully swallowed by a big bush. It seemed so unreal to glimpse the upside down car. There being no cell phones in those days, a couple of guys extracted me from the bush, carried me to a car and deposited me upright in the back seat for the trip to the hospital. I had broken ribs, a broken collar bone and a concussion. I lived.

My father had his beloved, mangled MG delivered to our place of business where he mounted it out front of our place on public display for all to see what his son had done. It wasn't quite at the level of a parent displaying the yellowed sheets of a bed wetter, but it ranked.

Any reasonable person would know that on the day of the accident I should have used my x-ray vision to have seen through the rock mass obstructing my view at the corner of the highway to observe that the oncoming car was in my lane.

With my ribs securely bound and my arm in a sling, the hospital released me the next day. Concussions were not dealt with in those days. We either died or went crazy. The good news, although I was completely unaware of it at the time, was that the laws of statistical probability suggested my chances of a second crash were now less. That included the chance of my crashing in an airplane.

My parents operated a motel in the Sault. They, especially my father, were active in the business community. The glittering MG, although not radically expensive, possessed enough local exotic image to serve well in the community as a symbol of business success then. Its exotic quality beat out the more expensive, but mundane, Cadillacs and Lincolns. Bimmers, Mercedes, Jaguars and such existed only on Mars.

So damage to the symbol was an event truly to be mourned. It lay prominently in state at the front of the parking lot. Each day that week Sault business friends, with mournful faces, gloomy eyes, suitably grey attire and hushed voices, came to the motel as if paying respects at a funeral parlour. It did not seem unreasonable for me to imagine the MG being placed in a prestigious coffin for burial, rites and all, but with me being carted off ignominiously to the wrecker's compactor.

I've related the MGA story because of its relevance in launching my aviation career. The car crash motivated my father to invest in a million dollar insurance policy on my life, and then to sign me up for flying lessons. A million dollars then was worth, in relative terms (No pun intended), much, much more than now, although presently, I would not discard such a gift.

It wasn't that my father didn't love me. His reasoning, based on his sincere beliefs, was that if the boy was going to perish young anyway, why not benefit his loved ones? In his mind, the car crash, even if I was innocent of malevolence, was a sign of the soon to be realized inevitable. My flying a plane would probably only hurry along the process a mere smidge.

So father, with the son who had experienced his second flight when catapulted from the rolling MG, headed together out to the new Algoma Flying Club to launch the boy's expectedly oh so brief aviation career. It's not true that my father's only thought during the drive was of himself seated in a newly acquired Rolls Royce Phantom parked in front of the first of his chain of luxury hotels.

Note: My father would have liked the parody.

Chapter 2

The First of Three Ways to Die

I'll be getting to the Embry Riddle aeronautical college experience in a bit, but first, I want to describe my first flying solo and a couple of other flights from that summer because it will provide a context to explain some technical points, and knowledge of my experiences could also save some newbie pilot's life. That's worth it all by itself. With my having been so relatively untrained as a pilot, I could easily have died three different ways during that first solo.

The Algoma Flying Club, before its relocation from the St. Mary's River to the soon to be built international airport, had no ground school. Printed aeronautical material and charts were available for purchase. We were to read the material on our own and simply ask the flying instructor questions about it if we wished. Otherwise, things would be explained in an ad hoc fashion as the 'lessons' went along. Basically, it was going to be a seat-of-the-pants, floatplane, bush flying experience.

The one instructor did not speak much. In fact, he seemed intimidating. I was reluctant to ask him questions, so I mainly just sat and observed. The instructor was a middle-aged man. I guess things hadn't gone well for him. He may have wanted to be an airline pilot, or own his own flying business, or some other business, but not be there earning a very low wage training newbies and not having the patience for it.

As for pilots aspiring to fly with the airlines, I knew one well known bush pilot who had flown transport planes for many, many hours during the Second World War. After the war, Trans Canada Airlines, as it was known then, asked him to fly international routes for them. He replied that he had absolutely no interest in sitting for all those boring hours at the controls (Personally, I could not imagine being bored with it). He loved the outdoors. He was resourceful enough to earn a fine living and raising a loving family flying his own bush plane, building to a prosperous fleet.

So I, simply observed what this instructor was doing during our flights, read the available flying material, and stayed mute most of the time.

The flying itself, especially flying off the river, was, naturally, exotic for a seventeen year old. My first landing was terrifying, but it soon became almost automatic for such young reflexes, although, an adrenaline rush did occur each time. That was wonderful and addictive. Unfortunately, I had joined the flying club late in the summer. By the time the operation was to be shut down for the winter, I could fly competently, but the instructor had stretched the dual flying time past the point where I could reasonably solo. I guess he just didn't want to chance the aggravation of a newbie crashing.

*

The winter passed. We now had a new flight instructor, another middle-aged man, newly arrived from Britain. I was thankful to see that he was very friendly and enthusiastic. For our first and only session together, I did all the flying, just circuits and bumps. He then told me that I would solo the next weekend. He asked me no questions about my flying knowledge.

Showtime came early the following Saturday morning. Mine would be the first flight of the day. I was surprised that so many of the other fliers were there, and there so early, just shortly after dawn. The mist was still rising off the river. And then I knew that I was to be the spectacle of the day.

This was as if a Roman Circus event. Word had gotten around that this was to be the young newbie's first solo. Will he freeze up and misjudge his first landing? Will the river swallow him? Only the bottom of the plane's floats would be visible above the water. It was as if the lions waiting below the ancient Roman Coliseum's floor were prowling back and forth restlessly. They could hear the babble of the crowd growing excited in the lusty anticipation of blood. They almost got their wish.

The club members arranged chairs in front of the very large clubhouse window, with its expansive view of the river, as if settling in to watch a movie. My nervousness, especially with knowing my every move was being scrutinized by the coliseum hoard, grew quickly as I walked toward the J3 Cub tied to the dock. I would have felt better if the instructor had accompanied me to the dock while uttering assurances that I possessed the skills for a successful flight, but he chose to stay in the clubhouse with the others.

Trying to focus on the task at hand, I noticed that the J3's fuel tank was full. The fuel indicator was simply a wire projecting through the fuel cap. A float attached to the wire at the bottom end rose up and down with the fuel level. The wire would bob up and down a lot during the flight, but it was reliable in its simplicity, with no electrical signal or other moving parts to complicate it.

The river's flow was undetectable. It was mirror smooth, with a rising mist. A loon cast a wakeup call to its buddies. That scene should normally have induced a wonderful tranquility in me, but I remembered the hyenas in the clubhouse. Okay, I'm mixing my metaphors. One moment they're lions, another they're hyenas, another they're a human crowd, but all were lusting for blood. Then I even suspected the loon.

It being the first flight of the day for the plane, I at least remembered to pump out the accumulated water from the leaky floats. And I did not fall into the river like some claimed later in their story's embellishment about the occasion. The task soothed my nerves somewhat. It delayed the actual flying, but, upon completing the pumping task, I now had to shift to flying mode. My brain froze. Now was only instinct.

It was now as if all rational thought was impossible. The instructor from the previous year had not taught me how to pre-flight the aircraft. I had simply observed the process. This new instructor must have assumed I had been trained properly. That was not a particularly bright assumption. My brain freeze made it worse.

A task checklist would have been nice, but neither the previous instructor nor the present one thought of it, and certainly a seventeen year old newbie never thought to prepare one. All was to be done from memory.

An external pre-flight check of the aircraft should have ensured that all visible hardware was in safety. I should have moved the control surfaces by hand through their full travel to feel for looseness and to listen for suspicious sounds. I should have checked the engine's oil level and looked for any signs of fluid leakage. I should have checked the propeller blades for any signs of cracking and, with fuel and ignition off, rotated the prop to check for looseness and more suspicious sounds like grinding. It would also have provided that very small spurt of oil to the engine's innards to lessen component wear upon startup.

However, having noticed already that the fuel tank was full, I simply untied the ropes securing the plane to the dock, pushed the plane out onto the river and hopped onto the passenger side float.

I then opened the plane's door to reach in and quickly set up the controls for the start. However, to my surprise, I noticed that the controls were already set. That ordinarily should have set off raging alarm bells in an experienced pilot's mind, but I, of course, was a naïve newbie.

The fuel shutoff switch was at On, and the ignition switch was set to Both magnetos. I gave the fuel primer a couple of pumps, but, distracted, I neglected to check the throttle position, another rookie mistake. Ordinarily, I would have advanced the throttle a touch for the start.

I stepped forward on the float. The plane had no starter, so I placed my hand on the propeller blade to start the engine. After my hefty downward thrust of the blade, the engine roared to life. The throttle had been left in an advanced position. The initial rise of the J3's nose as it wallowed in the water to begin a takeoff run catapulted me backward, crashing me against the wing strut. I held on. Fortunately, my seventeen year old, very fit body was lithely able to swing around to quickly wrench the plane's door open. I reached in to retard the throttle. The nose settled back down. The earth continued in its orbit.

With the roar of the engine, I naturally couldn't hear the collective shout from the coliseum crowd in the clubhouse, but my quick glance in their direction revealed faces plastered to the large window with other faces bobbing up and down behind them. That was an event for which admission should have been charged.

The instructor should have provided me with a printed pre-start check list, of course. And it would have been nice of him to have guided me through the pre-start check himself before my arrival to ensure the rookie avoided that life-threatening circumstance.

I could have died two different ways there. One is that I could have hit my head against the wing strut or the float, been knocked unconscious, and drowned.

The other path to doom would have occurred right at the dock if I again had neglected to check the cockpit controls but decided to rotate the prop for that spurt of lubricating oil. The engine could very easily have roared to life. I would initially have been rocked backward toward the wing strut, but then, as the aircraft's forward motion was suddenly restrained by the ropes securing it to the dock, I would have, in reaction, been catapulted forward with my head and neck entering the prop arc. The decapitation, right in view of the clubhouse crowd, would indeed have satisfied their grisly blood lust.

With the engine safely at idle, I stepped into the cockpit. My traumatized, impaired brain followed reluctantly. Unknown to me, there would very soon follow two more glorious opportunities for doom.

Chapter 3

The Second and Third of Three Ways to Die

My brain was merely frozen before. Now it was dry ice as the J3 crawled very slowly at idle toward a point on the river I thought should be the point from which to initiate a takeoff. I tried to focus. What was I supposed to do before takeoff?

Nothing came to my unsettled mind except advancement of the throttle. I knew there were other things to do, but I could only just stare straight ahead. Giving up the effort to focus allowed one of the rare, spontaneous, mumbled comments by the instructor of the previous year to slip through to my consciousness. "Wait 'till the needle's off the peg," he had uttered as he pointed to the engine oil temperature gauge.

I observed on this very cool morning that the needle was not yet off the peg. That allowed me more time to think as the plane idled along. A gust of wind whipped up a spray from a wave that caused the plane to shudder as the water droplets entered the prop arc. It then occurred to me that the gust came directly across river from the Michigan side. Another instruction filtered through. "Take off directly into the wind. Crosswind takeoffs are too tricky for a puke like you."

I wondered if that previous instructor had instructed me more than I thought. Maybe I, a mere teenaged puke, had not listened. Later, I would know that warming the engine up to the point when the oil temperature needle had just unpegged was insufficient. Warm-ups should wait until the needle is past the yellow caution arc and into the green arc of the instrument before a takeoff is commenced. The oil needs to be warm enough, therefore thin enough, to flow easily for correct engine lubrication.

Maybe that previous instructor was simply unaware that the needle actually needed to be in the green arc. Or maybe he didn't care that, otherwise, the extra wear would lead to increased oil consumption and the expense of a premature engine overhaul.

That instructor was old enough to have been a World War II pilot. Maybe he had been taught to just get the needle off the peg so he could scramble his fighter into the air quickly. Pilots in other fighters were coming with an intense desire to fill him with holes if he proved too slow. Time between engine overhaul had not mattered much, or not at all, in that situation. Where was he now? Maybe he was at the bottom of a lake, having arrived there after his oil abused engine prematurely quit while he was flying over a fog. Alternatively, maybe one of his flying students was the one at the bottom of the lake as a consequence of his "needle off the peg" instruction.

Another technical matter that should have been included that morning, and that I definitely was not taught back then, is that I was supposed to have observed the oil pressure gauge thirty seconds after starting the engine to ensure the needle was in the green. If not, the engine had to be shut down so that the lack of lubrication would not bring the engine to a sudden, very expensive, shuddering, screeching, terminal halt. That should have been another item in the non-existent procedure review the new instructor should have examined me on before the flight.

Because of the crosswind, I turned the plane around and headed back toward the clubhouse. There, the river was wide enough for me to easily take off directly into the wind. While I plodded on with the taxiing, nervously conscious that I was maybe boring the clubhouse crowd waiting for action, I did remember to perform a separate check of each of the two magnetos to ensure the ignition was working properly. I rotated the ignition switch from Both mags to the Right mag and observed an expected drop of around a hundred RPM, and then did the same with the Left mag with the same observed result. I then switched back to Both mags.

If the reader is wondering what mags (magnetos) are, and why I was checking them like that, I'll attach a technical section for more in-depth detail when I've completed the first solo story. If the reader wants only the story, the more involved technical description can be easily skipped. For now, a technically informed person may have noticed a flaw in my mag check performance. After checking the Right mag, I should have switched back to the Both position of the switch before going to the Left position so as to clear the fowled spark plugs, the ones not firing. The inoperative plugs fowl that quickly during a check, causing an inaccurate reading.

Some pilots believe it's only enough to observe any RPM drop at all during a mag check. I'll say briefly that it not only matters how far the RPM drops, but also how fast the drop occurs. Again, I'll go into more detail presently. I'll say more about the oil system and gauges too. Stay tuned.

Meanwhile, I tried to think of what else I was supposed to check before takeoff. The J3 did not have any flaps to set prior to takeoff. The propeller was of the simple but inefficient, fixed pitch angle design, so I did not have a prop control to operate to see if the blades variable angle function was operating correctly.

I should also have checked for freedom of movement of the flight controls, but did not. Many pilots check that the flight controls also move in the correct directions, for example, the left aileron rotates up to lower the left wing for a left (port) turn.

That directional check is absolutely necessary directly after the controls have been rigged or re-rigged. Such rigging would normally only occur during an inspection or other maintenance, but I suppose it's possible a maniac could sneak in and reverse the controls for some sinister reason. Unlikely, you say. Yes, but it has been known to happen. Some say that performing the directional check should be done automatically each flight simply as a matter of a safe practice that leads one to take other safe practices seriously. Some say it's evidence of a paranoid mind. Personally, I think paranoid pilots live longer.

Upon reaching the new takeoff point just in front of the clubhouse dock, my mind proved incapable of focusing on anything but the actual takeoff. Throttling down, I tried letting the plane naturally weathercock into the wind. That would help reduce my anxiety a bit by taking the blood thirsty observers behind the clubhouse window out of my peripheral vision.

The rising wind blew the morning haze away and initiated some small waves. I should have recognized that the wind should be turning me directly into the wind at a faster rate than it was. The fact was that there remained two important checklist items that remained unchecked.

One is that I was supposed to pull the lever to retract the water rudders, the small rudders attached to the rear of each float. If in the lowered position, they each operate automatically with flight rudder pedals movement to control the plane's direction on the water the same as a rudder on a boat.

Leaving those small water rudders down in the water could cause some problems. One is that the water resistance on the rudders would very much slow the weather cocking of the plane directly into the wind. That itself would not cause much harm, if at all, unless the plane was in a tight spot like a very small lake where a very short takeoff was needed. With the pilot having to guess the exact headwind direction, the inefficiency would be in a probable somewhat crosswind component reducing lift.

The pilot would also have to fight any lake or river currents trying to take him or her off in an unanticipated direction. The extra increase in water friction would reduce the takeoff even more. Again, that might not mean much if the extra fuel consumption and increased takeoff distance due to the various frictions is not important.

Another problem is that with the water rudders in the air after takeoff, those rudders, although small, are substantially dual additive to the aeronautical effect of the flight rudder. That could lead to unintended flight control overcorrections.

However, the most important reason for lifting the water rudders upon reaching the takeoff point is what could easily happen upon performing a crosswind landing. The water rudders turn in the same direction as the flight rudder. Upon touchdown, the plane will suddenly swerve to one side. The wind-side wing tip will dig into the water followed by a catapulting crash. If we're lucky, we'll end up sitting atop the bottom side of a float with the rest of the plane upside down under the water surface, paddling with our hands for the shore. The fact that the water rudders were damaged at touchdown would be the least of our problems.

*

Finally lined up in what I estimated to be the headwind direction directly across the width of the river, I decided it was time to advance the throttle. I think it was amazing that my mind in automatic, monkey-instinct mode reminded me to look around out the windows to see if any other planes were coming in for a landing.

Such planes would usually belong to either of two local bushplane operations, either Sault Airways or AirDale. I'll discuss them much more in the second volume of this book, but for now, I'll simply say that it was strongly inadvisable to do something stupid to earn the dreadful wrath of those strutting men's men. Vengeance would be terribly swift. As I pointed out, even my monkey brain took notice of the danger.

The Ontario Department of Lands and Forests, later known as Natural Resources, also operated their float planes, the de Havilland Beavers and Otters, on the river, but usually further up river. Their pilots were just as tough, but their uniforms made them appear deceptively less ferocious.

Seeing no approaching aircraft, I considered the path clear for takeoff. The path wasn't exactly clear, and another dangerous checkpoint item was to be encountered, but I'll get to that shortly. For now, I advanced the throttle. Experienced pilots know that the throttle should be advanced gradually if takeoff distance allows. That's because the quickly increasing engine RPM needs to be supported by sufficient oil flow for both lube and the extra need for cooling. Too quick an acceleration leads to what we term shock cooling. That leads to metal cracking and other unkind things.

The piston engines are usually and mainly cooled by airflow over the cylinder fins, but the oil also absorbs a lot of heat. Also, too quick an RPM acceleration could imbalance the crankshaft balance, leading to vibration and premature overhaul. The more powerful engines, unlike the small Continental 65 horsepower on this J3, have dynamic balances on the crankshafts which could be even more easily damaged.

I had not been taught that back then, but I think I advanced the throttle at a reasonable rate. As I advanced the throttle, I also pulled the flight control, a stick in the J3, full back toward my lap. That was to get the aircraft up on what's called the step. When up on the step, the floats skim along the water's surface with much reduced friction.

At first, while trying to get the plane up on the step, the nose points quite high. All that can be seen at first is the sky. As the nose of my J3 lowered upon reaching the step, I noticed that the loon I had seen on the river earlier was directly in my path.

Holding his ground, he seemed to be trying to belligerently stare me down. The airplane was just another bird, albeit a relatively very big bird. I was invading his territory. As I speeded closer to him he suddenly and miraculously transformed into a chicken and prudently scampered to a safe place. Unknown to me at the moment, he wasn't about to let the incident go. He was counting on a reckoning later. Apparently, loons have pride too.

With my reaching the takeoff speed, I pulled back normally on the stick. To my surprise, the plane did not lift off the water as usual, so I let the speed build up some more and pulled back on the stick again. This time the J3 lifted off, but then fell out of the air back to the water. Now noticing that the Michigan shore was looming precariously closer, I considered aborting the takeoff to, perhaps, then attempt a crosswind takeoff down river. There was not enough room left to safely fall out of the air again and come to a stop.

However, with my seventeen year old brain's fontal cortex not yet grown to adult size, I left the throttle fully advanced and once again pulled back on the stick. I even said to my dangerously naïve self, "Go for it!" The Michigan trees, as now seen very large out the windshield, slumped with dismay, closed their eyes and braced for impact. The clubhouse crowd, with a direct view of the event, grew quietly excited with the anticipation of blood.

At the even faster speed, I performed a 'duck walk'. The duck walk was a way of breaking the floats away from the suction of the water for a shorter takeoff. My wagging the control stick quickly side to side lifted one float out of the water and then the other. This time the J3 lifted into the air and stayed there – barely. The traumatized Michigan trees breathed a sigh of relief as the plane's floats skimmed over, but oh so closely, to their tops.

A hill loomed ahead directly in my path. As the Michigan forest zoomed by directly beneath me, I instinctively, seat of the pants, felt that the plane would plummet if I attempted to bank for a turn. I would try turning at the last moment to avoid the collision, of course. Better it is to let down in the softer tree tops than achieving intimacy with solid rock. Now I knew how the loon felt.

It was then that I recognized the force I was applying to hold the control stick in place had been progressively increasing as the airspeed increased. During the peculiar calm that sometimes briefly occurs when we realize death is unavoidably immanent, it occurred to me that the stick resistance probably meant that the elevator trim could be the problem. I reached up to turn the trim crank.

Eureka, the J3 ascended joyously unfettered into the sky – my threatened demise denied, at least for the moment. The clubhouse crowd catapulted from their seats with a joyous cheer.

The J3 had not been purposely set up that way to give me opportunity for a learning experience, of course. What I think had happened was that some uninformed person had naively played with the controls. Maybe someone's child had been placed in the pilot's seat and had a fun time. I was obviously fortunate to survive the experience. And yes, it was definitely a learning experience.

In the full flush of an adrenaline rush, I joyously banked the plane and headed for my family's residence, happily unaware that another opportunity for doom, and another adrenaline rush, were to very soon make their presence known.

Leveling the plane at a thousand feet AGL (Above Ground Level), I headed directly for my family's place, the Sault's Lincoln Motel at the junction of the Trunk Road and Wellington Street where there was also an Esso gas station.

As I, the young fellow who could now fly an airplane on his own began proudly circling over the motel, my parents and little sister came out to wave up at me. That's when I naively banked the plane into a steeper turn so as to see the family better. They for sure saw me return a wave.

The fun over, I then tried to pull out of the turn by moving the control stick to the right. To my dismayed surprise, the wings did not budge at all. The turn remained constant. Had I forgotten some other pre-flight check? Nothing came to mind.

You might think I had already depleted my supply of adrenaline. The truth is that teenagers have a seemingly endless supply instantly available for all stimulating occasions, but this time it pumped out at a slightly reduced rate. I could better comprehend the situation I was in.

The airspeed was dropping. I would very soon enter a death spiral. If I had to choose between crashing into my family or into the Esso station with an enormous explosion, I would naturally choose the enormous explosion. On the bright side, there would be two benefits. My demise would be without suffering, and it would eliminate the cost of cremation. Justice would finally be served for my neglecting the pre-flight check.

I'm here alive to write this story, so what happened? Did I somehow survive a crash? Was I suddenly beamed up into an alien pod? What happened is that the reduced adrenaline flow to my pre-frontal cortex challenged mind actually allowed me to attempt a rational solution to the problem, as in that peaceful moment of pre-death lucidity I had just experienced over the Michigan countryside.

What options were available to me? Advancing the throttle did not alter the turn at all. It probably only staved off the otherwise inevitable for a few seconds. Adjusting the trim wouldn't help here. Pulling back on the control stick would only tighten the turn and hurry the spiral. It then occurred to me that I had used the rudder pedals to get into the turn. It made sense to at least try to use the rudder to get out of the turn. I stomped on the starboard rudder pedal.

The wings leveled. Now a touch more wise in the ways of the world, and airplanes in particular, I very cautiously headed directly back to the clubhouse. The point to make with this last incident is that all student pilots, inexperienced as they are, should be instructed about extreme attitudes, challenging bank angles, before their first solo, and strongly reminded of it before that flight. So many have perished exactly that way. Maybe the very heavy floats caused a peculiar aerodynamic effect because of the extra centrifugal force caused by the heavy floats. Whatever, I'm thankful for the J3's very generous aerodynamic envelope.

The gradual landing on the river went well. As I taxied to the dock, a good end of flight procedure would have been to perform a live magneto check, but I had no knowledge of that then, of course. The way to perform the check is to quickly rotate the ignition switch to Off and just as quickly rotate it back to Both with the engine RPM just above idle. The usual instruction about this is to watch for a certain amount of RPM drop before switching back to Both, but just listening for the engine to cut out is what I always do. There's less chance of a damaging backfire that way.

The reason for the live magneto check is to save the next pilot the chance of getting his or her hand chopped off or worse if the engine surprisingly starts while the pilot hand rotates the propeller as a part of the pre-start check.

I did, however, correctly starve the engine of fuel during shutdown by closing the fuel valve. As I stepped out of the plane onto the dock after ensuring the trim, the throttle and ignition switch would not threaten the life of the next naïve pilot, the loon swooped down for his revenge. Like a dive bomber, he released his lethal load of accumulated intestinal waste in my direction.

He missed, but his backward glance cautioned me to at least think about the territorial integrity of my fellow living creatures. I hear crows have a similar, anal-retentive, obstinate obsession. As for the rest of my flying that summer, there would be three more precarious experiences to survive, and from which to learn.

Chapter 4

Technically Airplanes

This chapter will provide a more detailed explanation of certain aircraft technical matters touched on in the previous chapter, so if you simply want to read a continuation of the story with those three more precarious flying incidents, skip ahead to the next chapter.

However, if you're interested in knowing why an aircraft piston engine has two spark plugs for every cylinder but an automobile engine has only one spark plug per cylinder, why a pilot should observe the amount of an RPM drop during a magneto check, but also look for how fast the RPM drops, why magnetos are used for aircraft ignition but not used on cars, why an engine oil pressure gauge may deceive a pilot, and so on, keep reading.

Please note that some explanations will seem relatively simple to a technically sophisticated reader. My intention in doing so is to make some concepts understandable to a wider audience. However, other explanations, I hope, should challenge the more technically experienced pro.

Let's start with the spark plugs. There are two sparks per cylinder on an aircraft engine for safety reasons and regulations. If the ignition system fails to fire one cylinder's spark plug, the other spark plug on that same cylinder will, hopefully, operate. The basic idea is that we prefer that airplanes not fall out of the air onto our house. For that reason we need to have two completely separate ignition systems. The added complexity requires more maintenance and is expensive.

If an automobile engine ignition system fails, we are, hopefully, already on the ground. It's annoying, but we simply pull over to the side of the road and call or walk to somewhere for assistance. Cars sometimes do crash into houses, of course, which is also annoying, but they do not usually drop from the air, and it's not usually caused by ignition failure. So it's okay for cars to have the simpler, less expensive single ignition system with one spark plug per cylinder.

*

A person may wonder if the two spark plugs per cylinder cause the aircraft engine to produce twice the power of a similar, but one spark plug per cylinder automobile engine. The answer is no.

But one may point to what happens every time a pilot tests the two ignition systems before takeoff to suggest that there is indeed an increase in power when a second spark plug is added to a cylinder.

Operating a cockpit switch, the pilot turns off one system to ensure the other system is working, and then visa versa. When the one system is operating, and only one spark plug is firing per cylinder, a drop in engine RPM is noted. That naturally means there's less power. Moving the ignition switch back to the Both position in which the two plugs are firing again increases the power. That power increase is illusory.

One reason is that there's only so much energy in the fuel. When it's used up, it's gone. So okay, you say, for a given amount of fuel, why is there a loss of power when one of the spark plugs is turned off? All that's needed is for the fuel to be ignited, right?

The answer is how the spark plugs are situated in the cylinder. An automobile spark plug is located right in the middle at the top of the cylinder. It points directly downward at the middle of the piston. The two aircraft spark plugs are also at the cylinder's top, but are located next to the top's sides. They point down at the piston, but do so at an oblique angle.

When the two aircraft spark plugs fire together, their two flame fronts combine in the middle, which produces the same effect as the single, centrally located automobile plug. When the pilot turns off one of the aircraft plugs, there remains only one combustion flame front and it's at that oblique angle. As it expands at that angle, only part of it presses directly on the piston. The other part of the flame front wastes energy as it bounces off the cylinder wall. So that's why the pilot notices, and expects, the RPM drop.

So if one of the two ignition systems fails in flight, the aircraft will produce less power and fly more slowly. It is now not legal to be flying normally, so the pilot is required to land at the closest, safe aerodrome.

The aircraft will also not be able to fly as high as normally, so if the only alternate location means flying over a mountain range, one at least has the compensation of an added adrenaline rush to complement one's day.

*

Would the pilot notice that only one spark plug had failed in only one of the cylinders? It depends on the aircraft, its instrumentation, its power setting and the individual pilot. Power would be reduced in that one cylinder. That would cause an imbalance that would add to the normal imbalances among the cylinders of an engine. The imbalance would cause an extra vibration and a sound change that may or may not be detectable, depending on the sensitivity of the individual pilot.

More applied power causes more vibration. It's highly subjective, but it would be nice for a technician or the next pilot who's going to fly the airplane to know about it. That's why a pilot tests the ignition system before each and every flight.

The reduced power also causes reduced heat in the cylinder. If the only heat indication for the engine is an oil temperature gauge, I much doubt it would be noticed. A multiple cylinder head temperature gauge (CHT), which measures the temperature of all the engine's cylinders, would catch the temperature change on that one cylinder. But the pilot, or owner, or both, should be aware of what temperatures are normal for a given set of circumstances. The temperatures for each cylinder differ, and they get hotter toward the engine's rear.

If the techy wants to explore the heat detection method as a way of discovering the failed plug, he or she will have a pyrometer among his or her tools to determine which cylinder has run cooler, therefore has the failed plug. You may have heard the ugly rumor that one may spit on one's fingers to slap test the very hot cylinders after a run. All I'll say is to enter that minefield with extreme caution.

I recommend that a qualified techy be used to install such a CHT. The reason is that the installation is complex. Polarity must be observed, the leads must be of the correct length and in matched pairs because resistance changes with length, connections must be cleaned properly, and the hot junction must be torqued properly for the correct heat transfer. We can't just swap any harness with any gauge on any airplane. I'm sure you get the idea.

I'm sure that if the owner/pilot does decide to do the installation, not only will he or she follow the maintenance manual and product instructions precisely, but will use a torque wrench, a torque wrench that is calibrated. Some people claim that they only need an ordinary wrench because they can 'feel' the exact torque value in their innermost being. They claim to be super humans in that regard, however, their social skills are awful, so let's not envy their gift.

As for using the maintenance manual, it wasn't written for garage grunts. There are many things not included but assumed that the reader knows only through a lengthy learning and practice process. If unqualified owners/pilots insist on doing their own maintenance, please at least consult a qualified techy for insights. Inspections, alterations and repairs are supposed to be closely supervised and signed off by qualified people.

Also, make sure you use the 'visualization pause' before deciding to go ahead with your unqualified work. The usual vision I suggest is to imagine yourself flying over a seemingly never ending expanse of Florida everglades. Make sure you include lots of alligators, snakes and other awful beasties in the picture. Then envision the engine starting to sputter.

I should add that, unlike a multiple unit, a 'single' CHT is attached to only one of the engine's cylinders, the one that's normally the hottest. If, by luck, that's the cylinder with the failed spark plug, then we quickly identify the problem. The single CHT will nicely aid the pilot in not overheating the engine, but if a failed plug is on any of the other cylinders, the gauge is useless for that purpose.

*

What are magnetos, why are two of them on each aircraft piston engine, and why don't automobiles use them? As I mentioned above, piston engine aircraft, unlike with cars and trucks, are required to have two completely separate ignition systems for safety reasons. A magneto is a completely self-contained spark producer. It doesn't need a battery to initiate the electricity. Simply rotate its permanent magnet and the electrical signal is there.

It is also very compact so that two of them fit onto the back of an engine. Simply attach electrical leads to the distributor part of the magneto and viola, we have the two completely separate ignition systems.

The J3 club I flew for the first solo had no battery at all. I served the function of the battery. You remember that to start the engine, I had to stand outside the cockpit on the float to thrust the propeller blade down for a hand-prop start. I should mention, though, that the magnetos I was familiar with back then were very complex to set up and needed regular maintenance by a qualified techy familiar with the intricacies. It was not a task for the amateur garage grunt. Today's magnetos are simpler, sealed units, but yet still have enough complexity that they should still be installed by the qualified techy.

*

Returning to the magneto check I performed before the takeoff for the first solo flight, I did note that the RPM dropped, but I did not know that the drop should not exceed about 150 RPM for each magneto and that the drops should not vary more than about 50 RPM from each other. I did know that I was supposed to run the engine up to about 1200 RPM to perform the check, but did not know that the reason was to purposely stress the ignition system to make any faults more apparent.

I also did not know that I should have been watching for how quickly the RPM dropped during the check. Remember I pointed out that the pilot who flies the same plane regularly should be aware of what indications are normal for that particular plane. The pilot should then recognize if the RPM drop is either quicker or slower than normal. The reason is that the pilot can clue the qualified techy into the fact that a problem has developed with the ignition system that needs attention. It would be nice to avoid the engine quitting while flying over shark infested waters or while over a fog with treacherous terrain beneath it.

If the observed RPM drop is too quick, the most likely problem is faulty or fouled spark plugs. If the drop is too slow, the likely problem is that the ignition timing is out of adjustment, or that the valves need adjustment. Or there may be other reasons that involve more sophisticated diagnosis. That information is nice but not necessary for the pilot to know. The important point is to inform the techy that he or she needs to check it out.

If the problem is fouled spark plugs, for instance, the techy may advise us to alter our engine management. For example, the fouling could have been due to us using too rich or too lean a fuel/air mixture for certain flight conditions. The techy would discover that by removing the spark plugs and observing the color of the deposits. An engine always reveals evidence of our evil deeds.

*

A person may wonder if jet or turbine engines use spark plugs. The answer is yes, but there are differences from the ones used on piston engines, and there's one very dangerous reason why only a qualified techy should touch them on an engine.

More accurately, the sparking types are called igniters. The non-sparking types of plugs used on less powerful engines are called glow plugs. Unlike on piston engines, the plugs are only needed to ignite the fuel/air mixture, but are not needed to sustain combustion after that. However, the jet's inlet air may become turbulent under certain conditions causing a flame-out. So the pilot will usually keep the plugs energized during takeoff, landing and turbulence.

Igniters need a larger gap than piston spark plugs to ignite the much larger flow of air, therefore, require much more energy. That extra energy is what makes them dangerous. After engine shutdown, the immense charge still held on the system's large capacitor could easily, and has, electrocuted a person touching a plug. A prominent warning printed on the engine and in the maintenance manual advises the techy to wait for about an hour or more after engine shutdown before performing ignition maintenance.

Some of those igniters for the large turbofan engines are also radioactive too. It ionizes the spark path for better operation. They must be handled and disposed of in a very special way. Igniters are obviously beasties not to be trifled with.

*

In the second paragraph of this technical chapter, I mentioned that the oil pressure gauge of a piston engine may deceptively lie to the pilot. I'll conclude this chapter with an explanation of that.

Depending on the piston engine, the pickoff for measuring oil pressure may be located at either the oil pump end, which is at the beginning of the pressure system, or where the oil has passed through the engine, which is at the end of the pressure system. There are advantages and disadvantages to both systems.

If the pickoff is at the beginning of the system, just downstream from the pump, an advantage is that we know immediately about a loss of pressure from the pump. We can immediately shut the engine down and save a lot of money, which would be great if we survive the landing.

The disadvantage is that we won't know if some other kind of oil problem, like a leakage, developed downstream in the engine. In that case, the oil pressure gauge will read normally until it's too late for the engine. There's a time delay in the bleeding off of the pressure.

Conversely, if the gauge pickoff is at the end of the system, we know that pressure is maintained through the whole engine. If some kind of leakage did occur in a downstream location, the pressure gauge will sense that kind of failure in time so that we may hopefully save the engine. The disadvantage is that the pressure bleed off sensed at the gauge will usually be deviously slower. The pilot may not notice the problem until it's too late to save the engine. Choose your poison.

The pilot, therefore, should be including both the oil pressure and oil temperature gauges in his or her regular scan and comparing the two. With the end of system type, a rise of oil temperature will clue the pilot into the subtle pressure drop. Pilots should never ever be bored during a flight.

That's enough techyness for now. Let's get back to the story.

Chapter 5

The Fighter Pilot was a Normal Mortal

Located right next to the Algoma Flying Club that year when it was still on the Sault's St. Mary's River was a Norseman bush flying operation named Sault Airways. That Norseman was the same one that inspired my passion for aviation when it roared its six hundred horsepower Wasp engine to full power for the dramatic, water spray in my face takeoff.

The owner/pilot was the legendary bush pilot Keith Messenger, originally from Michigan. I do not believe in the ancient Greek gods, but he possessed that level of status because he had flown a fighter off an American aircraft carrier in the World War II Pacific struggle.

I often did chores for him that summer, like waxing his beautiful, white with yellow striped Norseman and so on. He was to have an enormous influence on my aviation career.

The crucial point in that influence occurred one day at his dock. One day, I waited there to help him dock the Norseman after a flight. As he came close to the dock, he cut the engine, quickly climbed out onto a float, but suddenly lost his footing and fell into the water. I grabbed the drifting plane to secure it to the dock.

Keith swam to the dock where he lethargically dragged himself up. The effort seemed to completely exhaust him. I asked if he was okay, but he just sat there mute. His breathing was laboured. With his wet hair flopped over his face, he looked like a drowned rat. That was disconcerting because a person possessing Greek god-like status was not supposed to ever look like a very mortal, drowned rat.

Maybe he was, gasp, a normal human being after all. But that couldn't be, I thought. Normal human beings could never fly a fighter off an aircraft carrier, and they could never fly a wonderful beyond words Norseman either. And they also could never ever feel discouraged. In my young mind, bush flying appeared too fantastic an occupation to have any room for such nonsense. But there he was. The situation did not look promising.

"You don't want to do this," he seemed to mumble dejectedly.

Was he speaking to me? I wondered. I thought he wasn't even aware of my presence. If those words had been directed to me, I must have heard him incorrectly, I decided. I waited, immobile and mute. Keith then lifted his head to fix me with baleful eyes. Yes, he was speaking to me.

"This flying stuff isn't as great as it looks to your young eyes," he added. He then went on to describe how labour intensive the work of bush flying was, which had become more difficult for him with age. He mentioned the long hours, the barely breaking even financially yet having to support a family.

He also mentioned that my finding work as a pilot in that era would be very difficult because pilots were a dime a dozen at that point after the war. The hordes of pilots still employed were in the middle of their careers then. They would not be retiring for quite a while yet.

A fellow my age then, he went on to advise, didn't stand a chance of making it was a pilot unless I had tons of money to start my own flying business or had influence somewhere. The timing was wrong for me. With sadness, I realized, at least temporarily, that I had been born in the wrong era to pursue that dream.

He had guessed correctly that I did not have tons of money, and did not know influential people who could help me in the aviation world.

"Get an education," he concluded.

No doubt his passion for aviation was diminishing, but mine was already steam-rolling. It was then that I recalled seeing an advertisement in 'Flying' magazine about the world renowned Embry Riddle, an aviation college affiliated with the U of Miami in Miami, Florida.

It mainly offered diploma programs in flying, engineering and technology. Degrees were bestowed by the U of Miami. Now it's still world renowned for excellence, but is a stand-alone university with its main campuses at Daytona Beach, Florida and in Arizona.

The program that interested me was Airframe and Powerplant Technology. A graduate would receive a technology diploma, but also qualify to write the exams for an FAA A&P technician licence.

My epiphany that arose from the ashes of no hope of a flying career suddenly morphed into a plan to succeed as a pilot, to not have to give up on that passion.

It occurred to me that a person with techy training at a world renowned aviation education, and with an FAA techy licence as well to add to pilot qualifications would stand a better chance of employment as a pilot than a pilot without it.

I already pictured myself bathing on the Riviera during a stopover as a very well paid Air Canada Captain. I could first apply as a flight engineer, the third person in the cockpit in charge of all the techy stuff. Pilot capable Flight engineers back then often moved up to the First Officer's seat, and then went on to the Captain's seat if competent. Pilots of the non-techy sort were more easily bypassed.

My parents very nicely agreed to the plan. Attending that college in Florida would be even cheaper than at a Canadian university then. The Canadian dollar was worth more than the American and American goods were a lot cheaper.

Selling the MG sports car would help with the cost. My parents were sure Miami had a fine public transit system. They were, of course, unaware of the coming Miami transit strike.

Chapter 6

Dark Vader and Another Way to Die

At the time my father paid up front for my flying lessons. He also paid in full for the training to acquire his own private pilot licence. However, within moments of him bouncing around during the first takeoff, his motion sensitive body converted his facial complexion to a sallow absence of colour and initiated a dreadful, triple orifice deposit on the seat, instrument panel and floor. The disgusted instructor vigorously advised him to either clean it up or get used to his head being mounted on a pike.

I then happily benefited by having the remainder of my father's non-refundable, paid-up flying hours awarded to me. I would have to use up the extra flying time before leaving for Embry Riddle at that exotic southern location in August of that year.

I did not receive any more flying instruction, but I practiced the basic skills I read about in the flying publications. The items for a pre-flight check were now indelibly branded on my automatic memory. I had already used up three of my lives. I hoped that I would be allotted at least a total of nine like a cat.

The flying club also had a Taylorcraft on floats, which was also fun to fly, but the club membership was growing quickly, so they also shortly bought a Fleet Canuck, a Canadian built, tube and fabric, side-by-side seated aircraft similar to the Taylorcraft. It would soon entertain me with certain survival challenges.

The morning the Canuck arrived at our club, the instructor briefly talked to the delivery pilot, took the new aircraft for a brief, single circuit flight, and taxied back toward the dock. I waited at the dock to help guide the plane in.

To my surprise, the instructor grinned goofily, hopped out and chirped, "It handled great. Take it for a spin, boy." He ambled toward the clubhouse without looking back.

Shouldn't I have some sort of familiarization before flying a different make of aircraft? I wondered, stunned. He probably would have pointed out that the Fleet Canuck was a plane not much different than the other two the club owned. It had an 85 horsepower Continental engine like the Taylorcraft, a stick for flight control like the J3, no flaps, a simple, fixed pitch propeller, water rudders, and so on the same. What's the problem? I guess I was meant to interpret the situation as a vote of confidence in my developed flying ability at that point.

So I proceeded with the pre-flight, this time without the anxiety of a blood thirsty mob observing me from the club window. The pre-flight did not reveal anything surprisingly different, so I hand-prop started the engine as with the others, climbed in and taxied toward a takeoff point. As I taxied, I cranked the elevator trim to its centre position plus one crank as an estimated takeoff setting.

I then noticed that the territorial loon seemed to be setting himself in a position directly upwind as if to intentionally provoke a confrontation. I was unaware then of how much damage a bird that size could actually do. If it slammed into the propeller, it could cause enough of a blade bend to seriously unbalance the prop. Especially at takeoff thrust, the severe imbalance could rip the whole engine from its mounts. The resulting centre of gravity change would plunge me straight into the river. Or if the bird slipped past the prop, it could bash straight through the windshield to nail my face with a similar result. This loon was not to be taken lightly.

After the warm-up and the mag check, I decided to leave the water rudders down as I had noticed Keith Messenger do when wanting to perform a directional change during the takeoff run. There wasn't much of a wind, so while the Canuck was in the wallow before the floats reached the step, I steered for a cross wind takeoff instead.

As the nose lowered on the step, allowing me to see the river, I noticed, as suspected, that the evil loon was frantically trying to scamper into my path. Was I supposed to feel proud that my brain had out witted a bird brain?

That contingency had distracted me from noticing that I had to exert some back pressure on the control stick to keep the plane level on the step. It needed more nose up trim, but certainly did not present the problem as with the J3 on the first solo when the trim had been set to full nose down. It did not raise in my mind any suspicions of what was to come that flight. The Canuck seemed to leave the water quicker than the other aircraft.

At altitude, I explored the Canuck's flight characteristics with both a power-on and idle stall, and gentle flight manoeuvers. In my chastened mind, I would never again purposely enter an unusual flight attitude.

After a bit more flying around, I decided to perform a touch and go on the river wherein I would keep the floats on the step, advance the throttle and take off again. Nearing base on the circuit pattern, I scanned the river below for the loon. To my relief, he wasn't there, but I kept vigilant as I turned onto base. As I turned to final, I cranked in what I anticipated would be approach trim.

Suddenly and alarmingly, a black Norseman swooped down from above like Batman with his cape spread out, looming very, very large directly in front of my windshield. While I had been preoccupied scanning for the loon, I had neglected to look for other flying traffic while I was in the circuit. If I had properly looked around, I'm sure I would have seen the Norseman on approach. The loon had almost won the grudge match with his sinister diversion tactic. Bird brains are not to be underestimated.

Now I was aware that I had messed with the world of Sonny Dale, the tough, man's man owner of the AirDale bush flying operation. I would henceforth have to continue looking over my shoulder for both Sonny and the loon.

AirDale was just a bit further upstream. His Norseman being black, and his reputation for toughness, lent a certain sinister aura to my teenaged impression of Sonny. The Batman metaphor seems inadequate. Batman did only virtuous things. Dark Vader is a better way to characterize Sonny's impression on my mind back then. My friend, Keith Messenger, who flew his bright, white and yellow Norseman was as if a Greek god, but Sonny with his all black Norseman, was Dark Vader.

With his faster airplane and coming from behind, Dark Vader must have observed me and probably decided to have some wicked fun by swooping down in front of me like that. I even imagined him cackling. The turbulence tossed me around in his wake. I lifted the nose and advanced the throttle to perform a go-around. This time, with my mind temporarily seized by an obsessive paranoia, I scanned constantly for both aircraft traffic and the loon.

I again entered final feeling reasonably sure there would be no further surprises. Mercifully, I was unaware that one more menacing surprise would present itself that flight. I decided to perform a full stop landing instead of a touch and go so as to taxi around for a bit to relax and collect the nerves.

My mind became immersed in the task as I flared for the landing. With the Canuck absolutely level, the floats kissed the water's surface with a pleasing, gentle skim. I expected the nose to soon rise as loss of lift caused the plane to surrender the step to wallow the floats down in the water.

Surprisingly, the nose dipped down instead. The horizon rose ominously in the windshield. I saw a lot more water than anticipated. The fear gripped me that the plane was about to flip over. There was no time to analyze what had gone wrong. I yanked the control stick back to arrest the rotation. Now I saw only water out the windshield. My breathing froze. My survival instinct allowed me to quickly unbuckle the seat belt and open the door for my escape.

Maybe it was the wind resistance against the open door, I don't know, but the Canuck's forward rotation arrested for a suspenseful moment and then, to the accompaniment of my exploding exhale of relief, the plane settled back safely on the floats. A wonderful, warm flood of buzz accompanied the adrenaline rush, but that was enough rush and buzz for now. I taxied toward the club's dock.

The instructor, grinning as if like an idiot, came down from the clubhouse to help guide the plane to the dock. I guessed he had observed the whole of the awkward events. Before I could step out of the plane, he said to me through the plane's open door, "Go back out for more circuits, boy. This time land with a slightly nose up attitude and progressively bring the stick back after touchdown as the airspeed decreases. I forgot to tell you that the floats are undersized, which causes the centre of gravity to be too far forward. They work fine on another airplane type, but got mixed up."

How could he forget to tell me that? I wondered with complete amazement. No doubt he was sending me back out to continue the flying before my mind totally froze up with an enduring loss of confidence, a variation of the get back on the horse after being bucked folk wisdom.

Looking back on it now, that airplane should have been grounded until outfitted with compatible floats. I continued flying for what remained of the hour with some competent, nose elevated landings. I guess it was the sharp instincts of a teenager that helped me adapt to the situation so quickly. This was true seat of the pants flying.

More fun times were soon coming that summer, and more opportunities for doom.

Chapter 7

The Hunchback and Another Way to Die

In addition to the Dark Vader, black Norseman, Sonny Dale operated a Cessna 180 and a Stinson Reliant float plane at his upriver AirDale operation. I remember that in the Trade-a-Plane publication from back then that the used Stinson Reliants, often called Stinson Gullwings, were priced very cheaply. I naturally wondered why a plane, about the same size as the wonderful, much more expensive, Norseman, and so beautiful to my young eyes of the era with its sensuous gull wing shaping, along with the powerful, snarling radial engine I also loved so much, went for that almost giveaway price.

And then I learned the reason. Those seductively curved gull wings had developed a habit of falling off when stressed, especially if the plane came equipped with the extra weight burden of floats.

Dark Vader's particular Stinson also had another catastrophe tempting feature. Attached to the engine was an oversized propeller. The prop's rotating blade tips, therefore, broke the sound barrier during every takeoff. A stop-action photo of those blade tips when subjected to that enormous, abnormal stress, would illustrate those tips bending back severely.

Therefore, metal fatigue would develop prematurely. Cracks would develop prematurely. Unless that vulnerable area of the blades was closely inspected after every flight, the tips could easily rip off. At takeoff RPM, the resulting imbalance of just one tip coming off would rip the engine from its mounts. The subsequent excessive shift of the Stinson's centre of gravity would then plunge the aircraft to the ground.

As you can imagine, the propeller tips breaking the sound barrier creates a very loud sound. In fact, it's a magnificent, snapping sound that was music to my inexperienced ears. I only heard the beauty of it. I did not then understand the danger, and I wonder if anyone at AirDale did either.

The rumor at the flying club was that an evil hunchback flew the Stinson. It fit in well with the black Norseman, Dark Vader image. I wondered if maybe it had been the hunchback who had threateningly plunged that Norseman down in front of me on that first Fleet Canuck flight.

We couldn't see the AirDale pilots getting into their aircraft because their flight operation was just out of sight beyond a point on the shore, so those pilots were left to our imaginations. Also, because of a sense of loyalty to our adjacent Greek god, carrier fighter pilot, all around good guy with the bright, white and yellow Norseman, Keith Messenger and his competition with Sonny Dale for the bush flying business, we did not commit the treason of visiting AirDale.

Every time we heard the Stinson's prop begin to snap on the takeoff run, we all rushed out of the clubhouse to see either the wings rip off or witness the engine rip off, followed by a plunge straight into the river. I even imagined a macabre cackle from the hunchback as he fearlessly stared down his demise. I suppose it was the monster within us all that anticipated the witnessing of such things. However, to the dismay of our inner monster, the Stinson always flew off safely to beyond the far horizon. The Stinson and Norseman can be seen today at the Sault's Canadian Bushplane Heritage Centre (CBHC).

*

One gloomy, overcast day that summer, I noticed a hole in the cloud cover. The sky beyond was an enticing bright blue. Unaware of the danger, I was drawn to it as if hearing a siren's song. I set the Fleet Canuck into a climb and headed directly for it.

Flying above the top of those brightly illuminated, fluffy clouds was truly transcendent. It was experiences such as that which addicted one to flying. Everyone else in the area was feeling the gloom of the day, but there I was escaping it. Endorphins flowed easily.

I flew around up there enjoying it all for most of my assigned flying time that session. Then I gazed around to again find that one hole in the clouds through which to descend back into the gloom. The hole must have filled in. Although I thought I had been flying in large circles, I didn't know exactly where I was. Earlier, the clouds had seemed high enough for me to not smack into a hill if I became lost in them.

The clubhouse stories about wings ripping off because of cloud lost, disoriented pilots getting into extreme aircraft attitudes leading to excessive airspeed now reverberated in my mind. I now had another opportunity for doom. Another adrenaline rush catapulted from its starting blocks.

Later, I would learn that the best way to descend through clouds with an aircraft that had such limited instrumentation was to set the trim and power for a gentle descent, select a compass direction and maintain that direction with the rudder pedals only. The descent would then be a gentle, safe, downward weave. One must disregard body sensations. They lie to us in that situation. Trust the instruments. An experienced scan would also include quick glimpses of the turn and bank and airspeed indicators.

The time to descend could not be put off any longer. The fluffy clouds lost their transcendence. The gloom below the clouds now did not seem so bad after all. I did set the trim and power for a desired gentle descent, but I would not be observing any instruments. I would trust my stupid body.

As the clouds, now malevolent, swallowed me, I saw only white out the windshield. My body did not sense anything unusual. My time in the clouds had probably only been seconds, but it seemed to go on endlessly.

Finally, I broke through the clouds' bottom, but the plane was in a nose down, extreme bank attitude. I don't know what the airspeed was, but it was gratifying to know that the wings had not yet ripped off. The Canuck was not yet in a spin. It was in a wide spiral that would soon lead to a spin.

With my having experienced the solo and been caught in an extreme bank situation when flying over my home, I quickly remembered to also use rudder to recover. The wings levelled. The nose pulled up into a safe cruise attitude. I was once again a touch wiser, but now had used up four of my lives.

Chapter 8

Pancake Bay and yet More Ways to Die

The morning before I was to leave for Florida for that highly anticipated aviation education adventure began very early for me. I still had to use up a few hours of pre-paid flying time left over from my father's allotment. It was good I would be flying all day so as to absorb my quickly mounting excitement.

As I arrived at the flying club, I noticed the two largest pilots set off from the dock in the side-by-side Taylorcraft. Both of them must have hovered around three hundred pounds. The plane groaned, and the floats seemed almost submerged. It was another occasion for spectator seating at the large clubhouse window.

With it being a very hot, humid, windless day, the aerodynamic forces on the wings would be considerably reduced. The humidity contributed to the reduced aerodynamics because the suspended water molecules were lighter than the air molecules. With no wind, the river's surface was very smooth, which maximized the friction on the floats. The loon prudently chose to not interfere with the spectacle. He sat perched on a tree branch, also with eyes intently fixed on the plane.

The Taylorcraft's Continental engine did have twenty more horsepower than the sixty-five horsepower J3 Cub, but even that seemed grossly inadequate for the situation. The heavy floats also added to the problem. The plane would have been overloaded even if it had been equipped with the much lighter wheels gear.

As the plane began the takeoff run, we sat silently in full expectation that there was no way that airplane would ascend into the air. And if it did, was it so overloaded that the wings would rip off? I think the instructor/manager should not have allowed this experiment.

As the groaning plane passed by the centre of our view, we noticed that the floats had only climbed only about half way onto the step as if mushing through the water. We saw the ailerons alternate their action as the pilot tried to relieve the suction on the floats with a duck walk. Sadly, the morose, groaning Taylorcraft was in no mood for a walk. The plane continued on and on until it disappeared beyond the point into unseen AirDale territory. We rushed out of the clubhouse to better see what would happen.

A few moments later, the Taylorcraft came back into view at the point, its hot engine still struggling at full throttle. The river water had been chopped up by the initial takeoff run, which reduced the friction, but the floats remained pathetically only half way onto the step. It was hopeless. We laughed as the two big guys finally pointed the tortured plane toward the dock to taxi in a plodding surrender. Considering what was to come that day, it was nice to have that comic relief. I proceeded to the Fleet Canuck to begin my day of flying. I wanted to make sure I used up all my father's remaining flying time.

Even though the air remained hot, and the wind had not yet picked up, the water was still choppy enough for a quick takeoff. I ascended to five thousand feet as I headed north to climb into very cool air. It was like having air conditioning. I visited friends who were camping at a couple of lakes, and then set off for Pancake Bay on Lake Superior to visit other friends. I would not be seeing them again until I returned from the college for the December break.

*

Pancake Bay is an Ontario provincial camping beach supervised by provincial park rangers. It's located about sixty miles north of the Sault. Flying at five thousand feet as I approached the bay, I again made a couple of rookie mistakes.

One is that I cut the engine to idle for a descent to the bay. Experienced pilots know that would cause too steep a temperature gradient as the cylinders cooled. It's often called shock cooling. That, in turn, could lead to cylinder cracks, spark plug fouling and other engine nasties occurring before the normal overhaul time.

A subsequent engine failure will only occur, of course, when we're flying over endless trees and rocks with large, hungry, territorial predators below us in the season of victual scarcity when they're especially paranoid about any perceived danger to their innocent cubs. The recommendation is for the pilot to apply partial power for a descent with a sensible temp gradient.

The second mistake was for me to neglect to apply carburetor heat for the descent to keep ice from forming in the carb's venturi. Pilots are cringing as they read this. Sorry. The urge to confess has overpowered my sense of pride.

The ice could have interfered with the mixing of the air and fuel in the carburetor for combustion. The moment of surprise would occur if I chose to advance the throttle for a go-around to abort a dangerous landing.

A person might observe that it was a very hot day. Ice would form? A venturi in a carburetor is a restriction in the air flow. The restriction increases the speed of the air inside the venturi. The speed increase reduces the air pressure there also. The reduced air pressure is supposed to suck the fuel into that turbulent air flow to atomize it for combustion. The problem is that, under certain conditions like the high humidity of this day, the reduced air pressure also sucks the heat out of the air, hence the ice.

Fortunately, with idiot's luck, the landing went well. Many campers came out in boats to greet my arrival. I felt like Captain Cook arriving for the first time at that tropical island.

After pulling up to the beach, I took my friends out for a ride. The tiny floats posed no problem at that point. My teen reflexes, that I thought would last forever, easily anticipated every flying nuance, as if the Canuck was an extension of my own body. I look back on it now with amazement as I awkwardly thrash about the virtual skies in my flight simulator program.

My friends had invited me to enjoy a tasty meal before my flight back to the Sault, so, while taxiing back to the beach after the final excursion circuit, I revved the engine to run the plane up onto the beach.

In performing that manoeuvre, I was blissfully unaware of the stresses to which I was subjecting both the floats and aircraft. Later, I would learn that the float attachment points received a strain that was further transmitted to the whole airframe. The corrosion protecting layers of the floats bottoms were abraded as if by sandpaper, and the floats rivets flattened slightly at their shanks. The floats sheet metal holes, although stronger than the rivets, may have elongated ever so slightly. That float stress and wear would cause the floats to gradually leak more than they already did anyway. The early morning pumping of the floats was definitely necessary.

As we sat down for the meal, the chief park ranger, a very wide man with height to match, came stomping my way. His intense, accusing eyes fixed directly on me. Today, I imagine him as a cross between Caesar and the character who played the lead in the movie Mall Cop.

"You the pilot?" he barked in a surprisingly high voice that was almost a falsetto. It took an immense effort for me to restrain a laugh. He stood in front of me with legs planted solidly in a wide stance as if ready for combat.

With a clenched jaw, he pointed out that no airplanes were allowed on the beach and that I could pay a hefty fine for doing such a bad thing, even though I was unaware it was a bad thing. I guess there was no excuse for not having read the provincial regulations, or for not leaving the plane in the water and tethering it by rope to a stone.

A big problem with the situation was that I had not yet recovered from the absurdity of hearing a child's voice coming from the menacing face of the chief. A laugh tried to gurgle up my throat, but I managed to strangle it. With what I hoped was a reasonably convincing approximation of a serious countenance, I responded with an apology, promised to respect the regulations henceforth, and told him that I was about to leave.

After the mini-drama of a long, silent stare, he said, "I'll let you off with just a warning this time." I imagined Caesar giving a thumbs up to spare a pathetic gladiator. He abruptly turned and strutted back up the beach.

My friends offered me a goodbye meal. While I had been eating, with my back to the lake, I had not noticed a peculiar characteristic of those Lake Superior waters. Even after a whole day of fine, windless weather, and there being no other kind of disturbance known that day, large, continuous waves suddenly emerged. Those waves had apparently developed very quickly while I had been absorbed in lively conversation and fine food.

Someone said it was pressure waves, but could not define how pressure caused it. Another said that it's caused by the movements of an enormous Lake Monster, that it's of the same species as the Loch Ness monster.

Then, as surely as night follows day, the stories of possible monster sightings started. I was accused of disturbing the monster by landing on its favourite bay. The monster was especially upset by my pulling the Canuck up onto the beach. It was a desecration. A possible fine was the least of my worries.

Whatever. The fact remained that as I was just about to head home, the sizeable waves were now crashing on the beach. With my having only flown off the river and small lakes, I had not yet encountered waves anywhere near that size. The small Canuck negotiating waves of that size seemed improbable. The flying club would consider me missing if I did not arrive back by closing time. They would send out the other two aircraft to search for me. Maybe even Keith Messenger, Dark Vader and the provincial air service would be put to the bother and expense of searching. They would not be amused.

Cell phones would not come into existence for decades yet. I asked if anyone knew if a phone was available in the area. Someone said there was one in the chief ranger's office. So okay, if the waves continued into the evening, I could prevent an embarrassing search for me, but I would have to confess to the chief ranger that the plane would remain on the beach for the night. And what would the monster think?

How long do pressure waves continue? I stupidly asked my companions. Naturally, a chorus of voices chanted that the waves have been known to continue like that for days and days and days. I then realized that I was supposed to leave for Florida tomorrow afternoon.

I mentioned that I wanted to at least try a takeoff. Realizing that a catastrophe was very much a possibility, my friends stopped joking. One suggested I wait a while longer. Another remarked, with all seriousness, that pressure waves like that always continue into the evening.

That did it. I resolved to try. They helped me turn the plane around and push it to the water. I climbed onto one of the bobbing floats, opened a door, reached in to set the controls, hand propped for a start, climbed in and tightened the safety belt. With the nose of the Canuck pointed directly into the waves, I tentatively advanced the throttle a bit in an attempt to taxi out to where the water was less shallow, to where the waves were consequently less severe.

The prop shuddered against the water spray. Quickly, the nose rose up. I saw only sky. The nose then dipped down on the back side of a wave. I saw only water. Even my yet undeveloped, go for it attitude brain, knew it was hopeless. As I tried to turn the nose back around, the starboard wing tip touched the water, but the plane remained upright as it wallowed crazily from side to side.

When the beach came into the windshield's view, I gunned the engine to slide back up on the beach. My eyes automatically and nervously scanned the beach for the chief ranger. It was with relief that I was spared that added stress, at least for the moment.

*

Near the time I was supposed to arrive back at the flying club, I decided, with dismay and a rising heartbeat, to venture over to the chief ranger's cabin to phone in my problem and to also contact my parents so they wouldn't worry.

The chief made me stew in my anxiety juices by not answering his cabin door for a bit. Maybe he was tending to other duties somewhere else at the moment, I thought. Finally, I heard the shuffle of heavy feet, and then the door creaked open. The chief stared at me like I was some sort of delicious insect caught in his web.

It occurred to me that if he denied me the phone privilege he would be the cause of the expensive search and rescue effort. I had at least a little bit of power to reduce the humiliation.

"I'm very sorry to disturb you, sir chief," I blurted, aware that my addressing him as 'sir chief' sounded silly. He seemed to like the expression, but there was no way I was going to kneel before him. "The high waves are preventing me from flying back to base," I blurted out. "A search and rescue effort will begin if I don't call in. May I please use your phone?"

He stared at me for a long moment as he envisioned explaining his neglect of mercy to his boss, the media and whomever important else that could crush his career. Realizing he was to be denied a carnivorous feast, he morphed from a predatory spider to a little boy deprived of an ice cream cone. He wordlessly opened the door and pointed to the phone on his desk with a gesture of futility. Was that a picture of Caesar on the wall?

I stayed standing to place the call so that I wouldn't violate his chair with my presence. I phoned the flying club first. The instructor answered with his usual friendly tone. After hearing my predicament, he replied that I was not to worry, that if I could not get back that evening, the waves would surely have subsided by morning. However, in the event that the situation persisted and I was late getting back, I would have to pay for a whole extra day of flying because others had booked flights in the Canuck. Knowing I would hopefully be leaving the country very soon, I agreed. My mom was glad to get a call and know what was happening.

The chief brightened a little when I thanked him profusely for use of the phone and his magnanimous generosity if he would allow me to sleep on the beach that evening. I could tell he enjoyed the effusiveness and the part about his being magnanimous. I did not raise the topic of the plane being back on the beach again.

*

A storm front rolled in that evening with much rain, wind and lightning. It passed quickly, but the beach was soaked, and it was then very cold, a sharp contrast that occurs very often in the north. The pressure waves persisted into the evening. When it became clear that there was no chance of my returning to the Sault, my friends graciously supplied me with a rubber ground mat and a sleeping bag. The pleasure of bedding down under the very bright stars was a sort of compensation for the adverse events of the day.

In my first dream, I found myself back in the chief ranger's office promising to pay the fine if he would call off the lake monster. With a sardonic grin, he doubled the fine for my sliding the plane back onto the beach.

Very quickly after my bleary waking at first light, I realized with alarm that the lake monster was still in a petulant mood. I also realized with much disappointment that it was not just another dream. The angry waves continued their relentless beating of the beach. My mind began adjusting itself to hopelessness.

While I was eating breakfast with my friends, my hope suddenly rose when I spotted what looked like a small, shallow sand bar toward the far south end of the bay that seemed not far off the beach. The sand bar would block the waves, leaving a smooth section of water that I could possibly use for a successful takeoff if there was enough room. I don't know why I hadn't seen it earlier.

One of my friends, John Nicolson, a future forestry PhD in the Sault, revealed that he had a rope in the car that could be tied to the tail of the plane. He offered to hold the rope to secure the Canuck in place just off the beach long enough for me to run up the engine before he released the rope for the takeoff run. That would hopefully shorten the takeoff enough that I would avoid crashing into the sand bar.

It was also another opportunity for doom, but I was certainly willing to give it a try. We all enthusiastically pushed the plane to the lake. With John in the water guiding the plane, we all proceeded down the beach. It would at least add more dramatic entertainment to the bay people's camping experience.

*

Arriving at the sand bar area, I observed that the water was indeed calm in the lee of the bar with a calm strip wide enough for the takeoff. It being long enough for a takeoff was a toss-up. As John tied the rope to the Canuck's tail, I noticed that the chief ranger was standing in his cabin doorway to observe and that most of the campers had emerged from their tents to witness this theatrical event. I said a goodbye to the grim faces of my friends. It was show time.

John grasped the rope firmly and braced himself with his legs planted widely on the edge of the beach. If I needed a short takeoff on land, I could have simply applied the parking brake, run the engine rpm up, and then released the brake to commence the roll. In this case, John's holding the plane close to the beach was supposed to act as my parking brake.

I hopped up onto a float, prepared the engine for a start, thrust the prop blade down and climbed in after the engine sputtered to life. I secured the seat belt tightly. There would be no magnetos test, of course, even though it would have been nice to know I had full power available to negotiate that sand bar.

With my being preoccupied with the task at hand, I did not think about the large audience waiting to witness the dramatic moment. Nor did I think about what could happen if I hit the sand bar. I supposed I would have either skidded off into the large waves beyond with a lot of rough ugliness but remaining upright, or the plane would nose over, burying me under the lake's surface. My use of the fifth life could very easily be the last life.

When the oil temp gauge needle unpegged, there was no more reason to wait. I opened the pilot's door and leaned out to give John the thumbs up as a signal that the fun was about to begin. He held on tightly and firmly braced himself.

My conscious mind may have concentrated on the task at hand, but my subconscious took in the whole picture, along with the very real possibility of catastrophe. Therefore, my heart began beating wildly as I quickly advanced the throttle. With the plane being in the shallow water so close to the beach, the wallow wasn't deep at all. The horizon remained in the windshield view.

To my surprise, the plane seemed to begin its forward motion the moment I had begun to advance the throttle, and the plane's acceleration seemed much slower than usual. As the sand bar loomed closely, I wondered with alarm if John had somehow become tangled up in the rope. Had I noticed in my peripheral vision that the rope was wound around his body when I leaned out to give him the thumb's up? Was he now body surfing behind the plane? Would he bounce off the sand bar, pulling the nose down into the water?

Or if he did clear the sand bar, I pictured him going for the ride of his quite possibly very short life. And the campers, seeing him trailing behind the plane like one of those long advertising banners, would get a full measure of entertainment. His landing would be a bit rough. Maybe I could lower him so very gently that he could water ski with his feet like I had seen in TV clips of the Florida water show people. The campers would break out with cheers and applause. All would end well.

It was too late to stop without hitting the sand bar followed by a probable nose over. I wagged the stick for two steps of the duck walk and yanked back on the control. The floats barely cleared the sand bar. After levelling briefly to gain airspeed, I banked sharply to look back. To my immense relief, John safely waved to me from the beach. The rope snaked harmlessly behind the plane. Absurdly, I looked for the lake monster to suddenly arise from the watery deep to swallow me whole. He or she must have been busy elsewhere.

Finally remembering to breathe, I continued the steep bank for the turn, intending a goodbye sweep past the observing campers and my friends on the beach. It was honestly not my intention, but as the chief ranger came into my windshield view, I discovered that the plane was flying directly at him. The chief's eyes flew widely open. He frantically dove into the cabin's open doorway as I veered past, continuing the turn to the south. I had come closer to the beach than I thought.

Even though I had come so close to the beach, I had still been over the lake water, so my being below a thousand feet did not violate the regulations. However, the incident did have the appearance of recklessness. He could still issue the fine for having the Canuck on the beach. I would send him a letter of apology before getting out of town and out of the country that very afternoon.

*

I believe four factors allowed me to get the plane off before hitting the sand bar. One was John's holding the plane with the rope for as long as he could. The wind and water blast against his face must have been immense. A compensation for him was the fairly unique, adrenaline rush experience. How many people have stood so closely directly behind an airplane for a takeoff?

Another reason was the cold, much less humid air that had rolled in with the stormy weather the previous evening. The heavier, more condensed air molecules no doubt contributed a much increased aerodynamic effect.

The two waddling steps of the duck walk broke the water friction, and the undersized floats, with their reduced surface against the water reduced the friction even more. They were the little floats that could, so to speak.

Leaving Pancake Bay, I hugged the shore line at a low altitude as I flew south because I was quite low in fuel. Soon, I passed by the Gros Cap bluffs and turned east to follow the channel leading to the Sault. I passed by the area on my port side where the Sault's international airport was very soon to be built.

The lochs came into view. A lake freighter was waiting for the water in the loch to be raised so that the freighter could gain access to Lake Superior. Was that how the lake monster made it into Superior?

A ferry was transferring people and cars across the Ste. Mary's River to Sault, Michigan. The international bridge was yet to be built.

At the Ministry of Lands and Forests aviation facility beside the river, the yellow with black stripe piston powered de Havilland Beaver and Otter bush planes were being prepared for their forestry patrol and firefighting flights. The name of the ministry would soon change to Natural Resources. The Beaver and Otter fleet would soon be augmented by more powerful turbine versions.

Further downstream, past the Sault's downtown, AirDale with its then to me, sinister black Norseman, Cessna 180 and precarious Stinson passed by. I did not see Dark Vader or the hunchback, but I was told that no one ever saw them in daylight. The day would come when I would see them both.

With the magnificent sight of Keith Messenger's bright Norseman beyond the point, I began a descent for a landing. But then I saw something very peculiar. With great surprise, I spotted a man in the middle of the river. He seemed to be sitting on the surface of the water. I levelled the plane for a fly over.

In a moment, I understood what had happened. The man in the middle of the river was a flying club member who was sitting on the almost submerged bottom side of one of the Taylorcraft's floats. Flying directly over, I saw the inverted outline of the submerged plane. The man must have flipped the plane over only moments before. A couple of men in the club's boat, one the frantic instructor, had just set out from the dock for a rescue. Naturally, I was glad to have been spared that outcome myself.

I landed further down river and taxied to the dock where a throng of members had gathered. Two of them helped me dock the plane. They confirmed that the accident had just happened. The unfortunate person was a student pilot who had botched his first solo landing. I wondered if the loon had anything to do with it.

The rebuilding of the Taylorcraft would be a flying club group effort supervised by Keith Messenger's mechanic. With all the commotion going on, I naturally did not try to talk of my adventures. I simply said a goodbye to some of those gathered, asked them to tell the instructor that I had arrived back with the Canuck intact. As I was about to leave for home, one of my club friends called out, "Sonny Dale's been looking for you." He had said it with a wry smile, so I took it as a joke. Or at least I hoped it was a joke. One must not ever mess with Dark Vader.

I naturally had not realized it on that day, but that floatplane, bushplane flying world as I knew it then on the St. Mary's River would change very much.

The Algoma Flying Club would very soon discontinue its floatplane operation and move to the new international airport. I suppose the club could have kept the float flying going on the river as well as at the airport, but I'm sure the Taylorcraft accident convinced them otherwise.

The Ontario Ministry of Lands and Forests with their Beavers and Otters would move their air service headquarters away from the Sault. Thankfully, the Sault facility on the river would become the Canadian Bushplane Heritage Centre, the CBHC, an excellent aviation museum and activity centre visited by aviation enthusiasts and tourists from around the world.

Keith Messenger would retire, but his Sault Airways facility would, over the decades, remain to gradually disintegrate as if into the earth itself, eventually not leaving any fossilized remains to prove that it had once existed there. The snarl of the Norseman's powerful engine on takeoff that had instantly invoked my infatuation for aircraft would exist only in my memory.

AirDale up river beyond the point would remain operative for decades yet. It too would eventually begin a period of disintegration, but not before my not too far off adventure there. Meanwhile, I headed home thinking only of the exciting adventure awaiting me at the college in that exotic, southern location, set to begin that very afternoon.

Chapter 9

Love at Second Roar and Embry Riddle

It was as if a late October arctic low had settled over the Sault on that August day. In preparing myself for the trip to Miami, I felt quite comfortable dressing in my best grey wool slacks. Nor did it feel uncomfortable or out of place to wear my winter long, cotton underwear. It was only natural to not want to have the wool of the pants to itch the sensitive skin of my legs.

Over a T-shirt, I put on a white dress shirt and tie. The burgundy tie matched my heavy wool, dark-blue blazer. The Algoma Flying Club crest sewn onto the upper pocket of the blazer would surely impress the Embry Riddle people who would, of course, be greeting me with the open arms of a very warm welcome. They would surely not laugh at the heavy wool of my dress pants and blue blazer in that tropical place.

I thought it would have been nice to fly directly to Miami, but Sault Ste. Marie, in both Ontario and across the river in Sault Michigan, had no airport that could handle an airliner. My parents drove me to a U.S. air force base a few miles south in Michigan. Upon arriving there, I seemed to sense that the people at the base gazed at me a touch longer than should have been normal. No doubt it was in admiration of my flying club crest, not the fact that my attire was winter robust enough to handle a sturdy blizzard. After all, they could not see my long underwear.

The flight to Chicago was in a DC-7. Fantastic. I loved that airplane. To my young, eighteen year old ears, the intoxicating growl of those four powerful, pounding engines was a magnificent symphony. Now I was glad I did not fly directly to Miami.

My disembarking from the DC-7 at Chicago and boarding a Delta Airlines DC-8 that summer of 1960 day was symbolic

56

of the change in aviation in the mid twentieth century – the transition from primarily piston engine airliners to primarily jet engined airliners with the jet's accompanying improvements in speed, reliability and safety (on average).

My ears were not impressed by the swoosh of the DC-8's turbine engines. I conjured the uninspiring image of the plane powered by four large vacuum cleaners. I craved the satisfying, pounding growl of the DC-7's Wright Turbo-Compound, piston engine airliners. My body wanted to again experience the intoxicating roar of that first love Norseman's powerful radial engine as I stood on that Sault dock the previous year. I wanted the pounding, rumbling roar as if fourteen hundred Harley-Davidson Fat Boys were at full throttle all at once. The DC-7's takeoff earlier that day had been very gratifying.

The Delta DC-8 had a roar on takeoff, of course, but it did not achieve that vibrating grasp of my bone marrow. I supposed I would eventually find a way of wrapping my heart around that jet sound, and I did.

*

If you haven't experienced a multiengine, piston engines rumble, you can approximate it by listening to binaural beats, or an album of beats that some people listen to for relaxation and mood elevation.

In pure binaural beats, two frequencies usually in the bass range are set at a small frequency difference. The mixing of the two frequencies produces a very low 'beat' frequency.

A beat of about 7 Hz is in the theta range. It approximates the rumble of a twin engine aircraft at cruise RPM. On my flight simulator, I set the engines RPM of a Beech Baron at a slight difference from each other to get that pleasing effect.

In a real airplane, one must be careful to not fall into the delta range with a beat of four to about a half Hz. Our brain produces a delta frequency when we're in deep sleep. I wonder how many accidents have occurred because of that.

Even then, we do not get the effect of a four engine propliner. In normal cruise operation, none of the engines of a twin or a four engined propliner turn at exactly the same RPM. A four engine propliner rumbles with a theta beat of about seven Hz superimposed on an average bass drone of the engines at cruise. It's intoxicating, especially with the added vibration that I don't experience with my simulator, although I believe chair pads have been developed to add it.

So why don't we just listen to a pure seven Hz without superimposing it on two or four bass frequencies? It's because our ears can't hear seven Hz. Our hearing normally doesn't kick in until about twenty Hz. And I like the bass effect as well.

Other than the airliner effect, I loved the discordant sound produced by a Cessna 337 as very delicious. That twin has the two engines mounted directly forward and aft on the fuselage center line. The mixing of the wind from the two props is wonderful.

I made it more wonderful by offsetting the two engine RPMs with the type of prop synchronizer that has a dial for precise teaming of the RPMs.

*

When the jetliner settled into its cruise, so very much higher than that of the DC-7, the flight attendant's (called stewardesses then) first order of business was to distribute to all passengers, without asking, a very thin packet containing two Marlborough cigarettes.

Cigarette puffing movie superstars and the robust Marlborough man notwithstanding, I had never even felt a temptation to smoke. I had been heavy into sports. If the coach proclaimed smoking evil, then it was evil. Half of our high school football team had been caught puffing by the coach and been exiled in disgrace.

However, as if by some magical emanation, that thin packet of evil on the tray directly in front of me thrust an illogically alluring pull on my impressionable senses. My hand, with a mind completely of its own, reached forward and picked up the packet. I nervously looking around me and froze when I saw someone who looked like the coach.

To my shocked surprise, the man lit up. I later learned that the coach had indeed been a closet smoker, but it wasn't him sitting there. Maybe it had been his evil twin. I then noticed that everyone I could see took out a cigarette at the same time and lit up. Many open flames in the cabin.

With trembling hands, I reached in and extracted one of the cigarettes. I put it unlit to my nose and breathed in deeply. The aroma was intoxicating. Delicious. The first whiff of the smoke wafting from the other passengers' lit weeds was even more intoxicating. Sweet.

Then an image of an older smoker I knew asserted itself. He had a chronic, hacking cough and suffered in a prolonged, agonizing way. His death came with an early stroke. I put the 'coffin nail' back in the packet and left it there on the tray. The intimidating image of the hacker won the mental struggle.

Later, I would learn of the problems tobacco smoke would make for aircraft, and the correlations for my body. Cigarette tar gummed up the pressurization outflow valves, causing a higher buildup of air pressure than intended as altitude increased. That strained the airframe – exaggerated structural weaknesses and reduced the life of the airframe.

The cabin air pressure would build to the point where the outflow valves would unstick and pop open, or the valves' pressure relief valves would pop. That not only popped ears, but added a further sudden stress to the airframe. That happened unless the pilots, swaggering, grizzled World War II vets who were chimneys themselves, stayed on top of the situation by manually, occasionally, opening the cockpit secondary pressurization outflow valve. It was a needle valve that was not so affected by the gummy buildup. When they opened the needle valve, the smoke came forward into the cockpit. That purged the cabin air for the passengers (Smokers apparently don't like second hand smoke either).

The pilots had to be aware of the first class passengers, though. Their clout could be with a sizeable bat. Usually their second hand smoke wafted back toward the rabble, the coughing, hacking ones who always survived the crashes.

Further smoking problems occurred for the air circulation system and some flight/engine instruments. Some of those very delicate instruments are vented to the cockpit. They easily start behaving erratically when tar coats their surfaces. It's not nice to crash into a mountain or a building in a fog, although it does have the advantage of low cortisone levels and an expedited passage to the hereafter.

Removing humidity from the air would also be a problem. The chemical coating of the coalescer which ordinarily extracted the humidity would be covered by tar and be very ineffective. If not kept clean, the cabin filled with fog. I've even seen a cloud layer form over the passenger's heads. I would, as well, mention about the thunder, lightning and hail, but I don't think you would believe me.

The outflow valves and coalescers needed daily cleaning in those days, especially when smog was factored in like in Los Angeles. Passenger travel for me back then was both fun and awful. I think my youth voted for the fun.

*

The DC-8 landed for a brief stop in Atlanta, Georgia. It was and is the Delta Airlines home base and main hub. A hot blast of one hundred degree Fahrenheit air assaulted my winter clothed body when I stepped out onto the air-stair platform. Some passengers disembarked and others came aboard. There were no enclosed loading platforms for connecting the airliner directly to a terminal gate in those days. A stewardess standing on the platform flashed her eyes at my woolens and visibly restrained a laugh. She had to depart my presence when I asked her if she knew what the temperature was in Miami. A ground cart hooked up to the plane to keep the air conditioning going.

We soon departed for Miami. And yes, it was even hotter in Miami, but what really impressed me was the horrendous summer humidity. I was also surprised by the darkening sky at that dinner time of day. Back in the Sault, the sun would not be setting until about nine-thirty PM – big difference. I had forgotten the school lessons about the tilt of the earth and the Tropic of Cancer.

The taxi driver said he knew where Embry Riddle was located. His sly, almost threatening demeanor prompted me to look around for other taxis. None were available, so I warily stepped in. We drove around for what seemed a long while. Was that the second or third time we passed by that downtown Miami hotel? The fellow was just making a living the best way he knew how, I guess.

To my surprise, he finally stopped the car on what seemed, in the now advanced darkness, a desolate road. Was this where he was going to do the nasty? I wondered with increased alarm. My heart beat increased to allegro. He pointed out the passenger side window. I saw a long, chain-link fence behind which was a long, squat, building. The bricks sported a diarrhea tan.

I asked the driver if he was sure this was Embry Riddle. I naturally expected to see a magnificent campus spread out before me. The tall fence had no gate in sight. This was ridiculous. Was I supposed to throw my suitcase over and then climb the fence? Was this just a joke on a young guy? I pictured him hyena laughing after leaving me there.

He pointed again toward the fence and said, "Crawl under the fence there." I saw a slight depression under the fence that looked way too shallow. "Knock on the door. It's the dormitory at the back of the campus. The Dorm Master will answer."

For the fare, the taxi driver demanded a ransom in gold bullion. I had no choice but to pay if I hoped to ever again see my luggage, which was being held hostage in the car's trunk. I suppose I was lucky to not find myself deposited on a pile of other bodies dumped in Biscayne Bay after my throat had been treated to the joy of a gushing surprise. I paid the ransom. He extracted the luggage from its prison, and then roared off to troll for another naïve sucker.

Dressed in my finest, and with eyes wary for throat-ripping guard dogs, I threw the luggage over and then scraped my way clumsily under the fence. I knocked on the only door and waited for the completion of the joke. No doubt a ragged, twitchy-eyed drug addict would answer in the hopes of receiving the latest drug shipment to this flop house. He would attack with an ice pick upon learning the truth and take my wallet.

I breathed deeply for my anticipated last few breaths of life with a beating, quivering heart. No doubt the taxi driver received regular kickbacks for the sucker delivery service. In the morning my cadaver would be on a table for practicing student surgeons. The taxi driver would get a share of that action too.

The sound of heavy clomps came from behind the door. Here comes the freak. The door swung open too quickly. Standing before me was a middle-aged, very muscular guy dressed in what appeared like military green pants and a T-shirt. The very close cropped hair and the heavy, shiny boots told me the story. Embry Riddle must have also been the name of a Miami marine's base. The taxi driver did indeed deliver me to Embry Riddle, it seemed.

The man's harsh countenance caused me to expect to hear a barked order compelling me to flop to the ground and give him fifty pushups. Either that, or I was about to be abruptly introduced to a shocking, terminal adventure. The taxi driver was probably circling around to get a share of that too.

But there came no order to drop for fifty. He scanned my winter woolens but maintained an expressionless, awkward silence. He did not seem especially impressed by my flying club crest. Supposing the onus was on me to explain my presence, I then half stammered my name and added that I was a new student showing up fresh from Canada to what I thought was supposed to be Embry Riddle, an aeronautical college. He nodded and replied that no such name was on his list for a dorm room. Ok, I supposed that this was indeed the correct Embry Riddle and that no evilness was going to happen. That was encouraging. I resumed breathing.

Thinking more clearly now, I assumed correctly that my parents had neglected to reserve me a dorm room. No big deal. I was eternally grateful to them for sending me to the college at all. I supposed that for this one night I could sleep under the stars with my luggage for a pillow. I certainly wouldn't freeze. But were hungry alligators prowling the neighborhood?

The Dorm Nanny went dryly on to offer me a one night's stay in a dorm room. Some new dorm students would not be arriving until the next day. I was to present myself promptly at fourteen hundred hours Zulu at his office in the main campus building.

That meant, as all aviators know, that the local time at Greenwich, England would be two o'clock in the afternoon Greenwich local time when Miami local time would be nine o'clock in the morning.

Some other new students would be arriving at that time who also did not have dorm reservations. Further accommodation instructions would be issued at that time. I almost felt compelled to smartly snap to attention, stamp my foot sharply, salute and bark, "Yes sir!"

As I was about to step inside, I swear I saw out of the corner of my eye an alligator trundle by with a limp Doberman pincer clutched between the beast's teeth. The dorm door clanked shut behind me as if in a prison. Nanny wordlessly directed me to my temporary dorm room.

*

The mattress was bare, and was made bleaker by walls of the same diarrhea tint as the outside bricks. The room's air was stagnant and oppressively hot, so I stripped to at least my cottons. I opened the curtainless window to a zero breeze and a symphony of cricket clatter. Eighteen-year-olds do not become exhausted, but I lay out on the bed anyway and, without a radio or TV available, gave free reign to my imagination. Teens then were more practiced at that than a twenty-first century teen.

Was that my imagination or was I beginning to hear what sounded like the beginnings of a long moan heard over the crickets' erotic chirpings? Was it coming from a human? Was someone sick or something weird going on? The moan became louder. With a panic building, I hopped out of bed and checked under the bed for a body. No body stared back at me. Whew! I stepped into the corridor. No moans, but I listened at several doors - no sounds at all. Back in my room, I went to the open window to look for a body laid out on the grass in some kind of grief. No body, at least one that I could see in the dim light cast from a street lamp, but the moan, even louder now, was definitely coming from outside.

Now the sound was more of a low pitched, growling roar than a moan. Was it an airliner? As the sound became much louder, I determined the roar definitely came from aircraft engines – large, powerful engines like hundreds of Harley Fat Boys at full throttle. I began to wonder if it was an airliner in trouble. With another panic building, I began to wonder if it would crash into the dorm. The trees across the road became progressively more back lit. It must be the landing lights illuminating the trees. It's coming straight at me, I thought with ever more bulging eyes.

A very large, four engined airliner, a Lockheed Constellation, appeared suddenly over the trees. It flew at what seemed like only a few feet over the dorm roof with a horrendous, blasting roar. At least I was going to live to see another day. I listened tensely for the crash. No crash came. The roar level now diminished, and continued to lessen as the seconds ticked on. Now another moan/growl was building, one that I sensed would again soon build to a horrendous roar.

It occurred to me that the dorm must be located only a very short distance, maybe a block or two, from the end of a runway. I was right. Embry Riddle and the dorm were located only a very short distance directly in line from the end of Miami International's runway 9/27. The taxi driver had taken me on a truly larcenous route through Miami.

A roaring DC-7 passed over the dorm next. Piston propliners had a much shallower takeoff trajectory than jetliners, and there were no noise abatement regulations in those days. I lay back down on the mattress. The roars were a tonic to my young years and psyche. That night was truly going to rock. What a great start to this tropical adventure.

Chapter 10

Bartholomew and Fadi

Upon my waking in the early morning, my attention focused immediately on the scents in that dorm room. The summer, tropical air surrounding me was so different from that back in the Sault. We often experienced high humidity around the great lakes, but this air was heavily mixed with sea salt. I would later learn how that salt content would greatly accelerate aircraft corrosion and how to control it.

The new scents included that of the salty air, but also that from tropical flora, emanations from ancient molds inhabiting the hidden areas of that World War II barrack, nearly fossilized rat droppings, accumulated layers of sleep breath from generations of soldiers and students gradually deposited on the walls over time becoming hardened in thin layers like shale in ancient sea beds, and maybe even alligator breath too.

I became aware that the cricket chirps no longer came through the open window. I guess they all found suitable mates and had settled down for a night of bliss. A faint roar came from a propliner taking off in the distance. The active runway at the airport must have changed. I heard some people, presumably other students, moving around.

I now assumed the other students would be wearing blue jeans and short sleeves, so I dressed that way. Upon exiting the dorm front door, I saw a spacious campus spread out before me with narrow, single storey, military style buildings, and what looked like a series of small aircraft hangars leading to a large, many storied building that looked more like a hotel than a college academic structure.

In fact, it had been the fine old Fritz Hotel on Miami's NW 36th Avenue just west of 27th Street. With a gleaming white façade, the main, tall portion of the structure had two arch-porticoed wings with fewer stories angled toward the street. Its impressive style and appearance would make for a very fashionable hotel today.

Built in the mid-1920s as a close hotel for the airport traffic, it encountered financial troubles and never did become a hotel. The Great Depression drove the last nail into the coffin of the hotel dream.

Sadly, and yet humorously, the building became home to thousands of chickens, usually forty thousand at a time. It became known as the million dollar hen house. Anything for a buck, or a chicken. The roosters were in their glory. They wore a mandatory tux at all times.

It was in 1940, with World War II looming and a need for trained aviators, that Embry Riddle moved into the building's south wing. Single story barracks and other related structures were built on the property behind. Soon taking over the whole building, the structure became known as the Aviation Building. At various times they shared the building with National Airlines, the National Weather Service and so on.

One may see the building by typing Fritz Hotel into a search engine. However, do not confuse the older building with the present Fritz Hotel in Miami Beach. One may wonder if there's a connection between the two Fritz Hotels. I'm not sure. All I'll say is that one should not be surprised when entering the present Fritz Hotel to encounter a haughty, strutting rooster in a tux.

The college stayed at that location until '65 when it moved to Florida's Daytona Beach with other campuses elsewhere. It became an international, highly recognized, stand-alone university. After Embry Riddle moved out of the building, it housed a clothing factory for a while, but was finally torn down in the late '70s, to be replaced by today's juvenile detention center and courthouse.

I firmly believe that the property's history as a hen house and its ultimate use dedicated to wayward youth, was not, and is not, symbolic of Embry Riddle's students as they were in the '60s of my era there.

*

I located a cafeteria in the large building, but checked my watch to make sure I had time for it. Nanny appeared to be one not to be messed with. His fourteen hundred hour Zulu command to be at his office must be obeyed.

The cafeteria was very busy with students. I got a tray, lined up to select a suitable breakfast, and then looked around for a vacant seat, spotting only one empty seat at a two person table.

The student in the other chair looked to be about three hundred pounds or more. As I approached, he did not look up. He just kept eating with fairly loud lip smacking sounds, like a beast fully engrossed in his latest kill. He only grunted when I politely asked to join him at the table.

He wore a dark blue suit that was at least of lighter summer material. A tie of the same dark color graced his white shirt. His attire was out of place here, as my winter woolens would have been. With a quick survey, I noted with relief that all the other students were dressed casually like me. Resigned to a breakfast without conversation, I simply began eating.

"Going to Africa," my table mate blurted out in a deep-south, American drawl, still without making eye contact. "Airborne missionary service. Need to learn how to fly them, fix them, and engineer spare parts from nothing. Embry Riddle seems the best."

His use of the words airborne and missionary in the same sentence was a bit confusing at first. Airborne sounds military, but the forces don't do missionary work. Deciding he was entering missionary service, I assumed his deep-south drawl was due to his coming from a Bible belt state like, perhaps, Georgia or the Carolinas.

Pleased to have some conversation, even without the eye contact, I tried to engage him the best way I could. I hoped he would not judge me for not going into missionary work or some human benefitting pursuit, and I wondered how the Africans would react to his lack of eye contact.

I thought it best to start with an introduction. "I'm Gary," I offered with an outstretched hand, "What's yours?" My hand must surely have been in his downward field of view.

He ignored my hand and my question. I must have looked like a dog pawing for a treat.

"You're a Canadian," he correctly observed with a wince.

The wince had been either a Freudian imperative or a reaction to bitterness in the food. I naturally wondered how a young man from one of the southern states could discriminate place of origin so well. Was I giving off some kind of scent? Did he have extrasensory perception and was this situation going to become even creepier?

"It's the way you pronounced the word yours," he revealed, to my surprise and interest, and still without looking at me. "I'm from Detroit. Meet lots of Canadian evangelicals from Windsor. They pronounce the ou as almost a hard o as in oars. We pronounce it with a soft e like in yers."

So he was from Michigan, not a southern state. The erroneous assumption I made was that all North American evangelicals spoke with a Bible belt drawl. I thought that if he was going to be in my program, he was going to be a very interesting part of the experience.

"Name's Bartholomew," he said, still without looking up.

"Should I call you Bart?" I naively asked.

He flicked his hand as if swatting a bothersome, unsophisticated fly. "Bartholomew, of course."

Okay, so Bartholomew he would henceforth be. After breakfast, I noticed it was almost time to show up at Nanny's office. Bartholomew rose up as I did, revealing that he was about six foot two or three. Maybe he was also going to Nanny's office. He followed wordlessly behind me.

*

I noticed that about a dozen other young fellows were gathered in Nanny's office when we arrived. Years later when I taught at a college myself, at least a few women would be among the students. The term 'non-traditional occupation' was yet to be devised.

All eyes except those of Nanny, who was seated at his desk reading from his clipboard, slanted automatically toward Bartholomew when we entered. Bartholomew found a corner and stood while keeping his head bowed. A complete and awkward silence ruled the room while we waited for whatever something from Nanny.

In retrospect, I believe Nanny deliberately used the tension of a protracted silence to insert military-like discipline and control. It was as though he would suddenly arise to vigorously impress on us that this was not going to be an airhead, university experience, that this was the beginning of a very, very serious aviation education, that human lives - those of ourselves, our parents, our siblings, our grandparents, our friends, our future mates and children and their progeny, and billions of others - would depend on the integrity of our learned decisions and our very hands themselves.

But that didn't happen. Nanny abruptly checked his watch, as if he magically knew the exact moment the watch's precise sweep seconds hand would bang on fourteen hundred hours Zulu. I felt, mixed with a bit of anxiety, relief that something was happening.

Nanny's eyes flitted for a long moment to the blue suited, downcast, Bartholomew in the corner, the only one wearing a suit. I congratulated myself for having made the right decision about my casual attire. Bartholomew did not seem to notice Nanny's rude stare. He remained gazing at the floor.

As if with an immense effort to divert his eyes from the big guy, Nanny looked us over in silence and began pacing back and forth, still in silence, while continuing to look us over. Again in retrospect, I understand that his intention was to further intimidate, to exert control. I expected him to suddenly bark out that we were puke nothings and to drop and give him fifty pushups for not snapping to attention at exactly the fourteen hundred hours Zulu. It was only my imaginative anxiety, of course.

When he did speak, though, it was with an utter deadpan seriousness. Without barking out the imagined speech about aviation seriousness outlined above, he called out each one of our names from the class list, observing each face after a "Present" response so as to put a face with the student's name.

I would soon be impressed by Nanny when I noticed that he seemed to know the name of every student who ever attended the college, where they came from, and personal details about them.

The only student who did not voice his presence during that roll call had the family name Fadi. Nanny stated that we were the only ones of the incoming class who were not registered at the college dormitory, and who had not otherwise indicated they had arranged for other accommodation.

Nanny then handed out an information sheet which noted various essentials like the location of our classes and which tools to purchase. The noted class starting time of seven-thirty in the morning impressed my eighteen year old mind as noteworthy. That departure from a normal high school starting time of nine o'clock was viewed by most of us with an internal sigh at the prospect of a teen circadian clock punch worthy of hell on earth.

The two in the afternoon finishing time for a nap would help a lot. One should note that Miami was in a tropical zone, that it was mid-summer, and that the classrooms were not air conditioned then. I would late encounter an equatorial zone where work began at five AM with a merciful finishing time of ten AM.

Nanny went on to state that we could all stay at the same hotel in downtown Miami, that we would get to know each other during the first week, naturally select compatible apartment mates, and probably rent accommodation together at the Silver Springs apartment/hotel complex beside the airport.

Nanny then directed us all to accompany him to a waiting bus. He would drive us this one time only to the downtown hotel. We could subsequently use Miami's fine public transportation system to get to the college for classes.

As we rose to leave, Nanny opened his office door revealing a very tall young man standing there with a large luggage bag draped over his shoulder. He wore a Texas style Stetson, a bright cowboy shirt with rhinestones and high heeled cowboy boots. The high heels projected his already tall frame to about six foot six or higher. At least his boots did not sport spurs.

"Fadi says howdy, y'all," he barked with a goofy grin that was undeniable evidence of his true inner self, as we were soon to discover.

Nanny, without saying a word, walked past him. We all followed Nanny, trying to ignore Fadi's retained idiot's grin. With an exaggerated swagger, Fadi followed behind the group and boarded the bus with us.

The hotel on Miami's downtown Flagler Street was well past its prime, but could not exactly be described as seedy. I later wondered if Nanny was getting a percentage reward for sending the students there, month after month, year after year.

The hotel manager, or whatever his function was, waited expectantly outside the hotel's main entrance. Nanny stopped the bus at the front of the hotel, remained in his seat, and simply told us to disembark.

Most of us exited promptly with our luggage. Fadi and Bartholomew remained sitting in their seats until Nanny hurled profanity at them as he directed them to get their butts moving. Bartholomew exited first, followed closely by Fadi, who seemed to have a gloating expression of victory.

In retrospect, I know what happened. Bartholomew, prisoner of his insecurities, very much had wanted to trail the group so he could securely judge what was coming. Maybe he had been bullied in earlier years and also did not want to have someone with malevolent intent behind him.

No doubt, Fadi, with much previous practice, sensed Bartholomew's insecurity and wanted to feed on it. It was natural that the emotionally weaker one would exit first, and promptly, at Nanny's order. Fadi would probably have been quite willing to face Nanny's wrath to achieve his goal.

The hotel manager led our group inside. He scooped up a handful of keys from a counter and said to us, with a Cuban accent, "You don't know each other. Two beds in a room. Two people to a room. I tell you who's with who and where. You don't like it, you can switch if anyone's willing or you're stuck. But you must tell me if you switch."

Everyone offered their silent obeisance. I reflected on the fact that this was not an army base barracks. We could simply find another hotel if we wanted. Conversely, Nanny may be upset, and maybe Nanny had arranged the best per Diem around. I would go with the flow.

A slender woman, perhaps of Cuban extraction, attractively middle aged, sauntered down the hotel staircase in as suave a manner as she could manage. She halted to look me over for a moment, and then continued out the hotel entrance. The guys looked at me with a mixture of surprise and awe. I just shrugged. Fadi quickly followed her.

The manager handed out keys to appointed pairs of us as we progressed to the higher floors. Finally, three of us were left on the very top floor, me, Bartholomew and a fellow called Texas. Would I be rooming with Bartholomew? I wondered with a building anxiety.

Before the manager could speak, we heard heavy clomping on the stairs. In a moment, a puffing Fadi made his appearance, his goofy grin resurrected the moment he saw us. Maybe it hadn't worked out well for him with the woman. Maybe she was a professional lady who had a rule about no business with neighbours. Now that Fadi had arrived, the prospect of rooming with him caused me even more anxiety.

Bartholomew's face reddened, and then he panicked. "I need to be alone," he blurted, almost shouting.

The manager stoically responded, "Two to a room. One will have to pay double if you leave. You want to do that to your friends?"

The appeal to conscience was a good ploy. The manager handed a key to me and indicated that I would be rooming with Texas. Bartholomew's eyes widened in terror as he realized he would be rooming with Fadi. Fadi grinned sadistically.

Chapter 11

The Boot Salesman, Sockets and Tiny

My new roomie, Texas, was a wirey, sturdy fellow probably a couple of years older than me who appeared as though he wouldn't take much nonsense from anyone. He was from a town bordering with Mexico. It was exotic to me to listen to his authentic, unFadi-like, Texan English with added Spanish words like mucho and gracia and so on. It was like me in the Sault inserting French words into my speech like Oui, merci, sa va, and so on similarly. We seemed to naturally lapse into more and more use of our second languages, explaining the meaning of phrases.

We got along very well, but he was taking aeronautical engineering and I was enrolled in technology. During that first week, it would be natural for the new students to team up with people from our own program as apartment mates for the rest of our stay at the college – engineers with engineers, pilots with pilots, techies with techies.

We heard that a hurricane, Donna, was immanent and headed for south Florida. The weather people predicted the eye of the storm to pass directly over Miami. As predicted, the winds came. As they rattled the windows, I envisioned a locomotive, or at least a coconut, to smash through the glass, inflicting an inglorious, but quick, end to my short life.

The hurricane's eye passed over us. Then, suddenly, all became quiet, sweetness and light. We were hungry. Naïve about the nature of hurricanes, and unaware of the true devastation around us, we walked to a store for food, enjoying more of the sweetness and the bright, blue light of the now cloudless sky.

The grocery store's windows were boarded over, but it was open. Business was being conducted with the usual joyous

Cuban attitude, almost as if nothing unusual had happened. We then headed back toward the hotel with our groceries that mostly included items that did not have to be cooked and that did not spoil too quickly.

The trailing edge of the hurricane passed over us as we headed back toward the hotel. The savage winds returned suddenly. We must have looked ridiculous clutching our grocery bags to our bellies while leaning into the wind at an exaggerated slant, like circus clowns do in their big shoes to get a laugh. In that lean, I could have easily reached out straight ahead of me to touch the ground.

In retrospect, I now know what danger we had been in. Lethal projectiles of one sort or another must surely have been launched in those ferocious winds, but I saw none. To me, and Texas as well, we were simply in the middle of a great adventure, naively oblivious of the surrounding damage to person and property. Back at the hotel, we heard through the wall Bartholomew ranting, almost screaming, at Fadi. I imagined Fadi's sardonic grin.

Bartholomew's ranting and shouting soon stopped, but now and then there what sounded like Fadi occasionally shouting, "Yip, yip, yip," sometimes followed by a disconcerting crash against the wall. It wouldn't be long before we found out what that was all about.

*

Then came the first day of instruction at the college. After the instructor introduced himself and his assistant, he set a small display stand of work boots on his desk. His jaw-clenched smile suggested, at least in my mind, that we should buy from him if we expected a good grade. Some bought. I did not, because I already had functional safety boots, and my funds were limited.

As would be the pattern, our classes began with a lecture followed by shop to apply the theory learned. That first day was aircraft stimulating. We cut and steamed strips of spruce to build wing ribs in a jig. Subsequent classes during that first

76

month included the various groups building a wing section covered by fabric to its sewn, primed, painted completion. Fabric and wood repairs were then to be performed on that section.

*

The second day was when the personal dramas began. I emphasize the word drama. Our boot salesman instructor began the class by glowering red-faced at us. Had his demonic posture been inspired by not enough of us buying his boots? With his clenched fists at the end of his ramrod straight arms, and his shoulders hunched, he resembled Bartholomew in an intense moment.

With cheek flapping fury, he angrily barked, "You complained to my boss that I was trying to sell you boots."

His eyes did not settle on any one person. We were all guilty whether actually complicit or not. I knew nothing of any complaint. Some of us had joked about the boots thing after the initial class, but that was all I knew of, and judging by the saucer eyes of the other students, they appeared to have no knowledge of any official complaint either.

"It's a disgusting lie," he shouted with a Bartholomew-like foot stamp. His demeanor now shifted from anger to a palms forward, quivering plead. "I was only showing you the kind of boots you should get. This has been a terrible, terrible insult to my person."

He stared at our incredulous faces silently for a long moment as if hopelessly waiting for some kind of statement of understanding from us that would never come. His face then reddened even more. His lips tightened. Suddenly, he pivoted and stomped out the open door.

We naturally remained seated in a stunned silence. The teaching assistant appeared to be wide-eyed stunned as well for a long moment, but then clasped his hands together and began lecturing as if nothing had happened. After his competent, interesting lecture, he set us up in the shop to assemble the formed ribs of the day before onto individual, short sets of spars for the wing sections we were to cover with fabric.

We first had to use the socket set from our newly purchased, gleaming set of tools to bolt the two spar pieces into a provided light, outer frame. Moments after we all began the task, the second unbelievable event of the day occurred.

Bartholomew suddenly screamed and fell to the floor, writhing around with teared eyes as if in intense pain. Was it kidney stones? Was he going to projectile vomit? I unconsciously stepped aside, out of his direct line of fire. Also unconsciously, it was as though my eighteen year old mind understood that it was normal for this unusual event to happen to a person like Bartholomew.

The teaching assistant rushed to him. Bartholomew tried to speak, but he had trouble getting breath for it, as though his diaphragm had seized.

The teaching assistant calmly said, "Whisper it to me." He then cocked his ear close to Bartholomew's quivering mouth to listen.

Bartholomew stopped his writhing and whispered some words to the assistant. In retrospect, I think that was very insightful for the teaching assistant to handle the situation in that un-panicked way. I wonder how the boot salesman would have handled it.

The assistant rose up and surveyed us with what seemed like a curious mixture of restrained mirth and yet accusation. "His five-eighths socket is missing," he announced finally as Bartholomew's massive hulk remained curled up on the floor. At least he wasn't sucking his thumb. "Does anyone, by chance, know what happened to it?"

I had to restrain myself from breaking out in rude laughter. I'm sure others were experiencing the same inner struggle. But then, more somberly now, it occurred to me that I should check my tool box. Except for Fadi, all the others did the same. Sure enough, a number of the five-eighths sockets needed for our project, including mine, had gone AWOL.

Naturally, most of us from the hotel suspected Fadi of having perpetrated the foul deed. All eyes turned to him. His goofy grin became even goofier, almost evil, as though he had gained a hefty load of sadistic pleasure in the attention.

When some of us mentioned that our socket was also missing, Bartholomew sat up with an expression like that of a little child who had been offered ice cream to draw him out of a tantrum.

"Strip-search us, sir," taunted Fadi.

All eyes turned to the assistant. How would he handle this? His mature handling of Bartholomew moments before offered hope of a dignified resolution.

"Okay, everyone," began the assistant, his expression now surveying us intently, as if having crossed the anger threshold, "Drop your pants."

We all laughed loudly. His comic relief approach to at least the initial stage of the problem was an excellent tactic. Bartholomew now rose to his feet. He scanned us with an expression as if he did not understand the humor.

The assistant spoke again. "If this is someone's silly joke, please cough up your booty now so we can all enjoy another laugh and then get on with the project."

All eyes again turned to Fadi. His grin widened to its maximum as he stood there silently with the outstretched arms of a person pleading innocence.

*

That's when Tiny stepped forward. Tiny, from a Wisconsin farm, was the biggest guy in the class, about two hundred forty pounds of linebacker-like muscle. His massive jaw jutted out beneath a marine style crew cut.

"I saw you slither back to the shop area when all the others were distracted by the weird instructor," accused Tiny, now almost touching Fadi's nose with his. "Empty your pockets, creep."

Fadi just stared at him silently with the same plastered grin, as if to proclaim total innocence, "What? What?" Fadi seemed about the same height as Tiny, but by the way he walked with a meandering teeter, we all believed Fadi had massive lifts in his cowboy boots.

The punch came quickly. Fadi catapulted back onto the floor. Gleaming, five-eighths sockets escaped his pockets and tried to escape like mice frantically scattering for the safety of hidden crevices. Fadi still retained the goofy grin. Some blood trickled from his left nostril. A black eye would soon develop.

The teaching assistant's eyes shot upwards, as if beseeching extraordinary guidance. Either that or it was the initial stage of a nervous breakdown. "Okay, okay," he began after bringing his hands together in the steeple design, his senses restored. He looked directly at Tiny whose expression was one of helplessly observing his beloved aviation career prospect departing for Neverland. "Let's salvage what we can here. You and the socket enthusiast will come with me for a discussion with Nanny. We'll try to work it out."

Fadi scrambled uncertainly to his feet, his intense eyes darting around like a cornered rodent arrogantly flashing the stolen cheese in his mouth.

The assistant turned to address the rest of us. "Y'all can continue with your project while we're gone. If you can't locate your socket, you can use your adjustable crescent wrench for this one time only. I'm sure our Mr. Fadi here, if he survives the Nanny interview, will reimburse all losses with interest. "

Suddenly, the assistant threw his head backward and shot out his arms stiffly. The shocking gesture even removed the goofy grin from Fadi.

The assistant relaxed his posture. His face then contorted into a sardonic grimace, as if Fadi's goofy grin had been transplanted directly onto the assistant's face. "I just removed the curse on all your crescent wrenches," he explained. We now knew he was trying to use comic relief to relax the tension. It worked. "The crescent wrench," he went on, "is only in your tool box to remind you to never ever use an imprecise tool when a more precise tool, like a socket, is available. The crescent wrench curse will be reinstated at the end of this day."

In our careers we would learn that precise tools would not be available, or even exist, for every application. We would, in fact, need to design and fabricate our own tools in certain situations more often than one would imagine. That was one of the reasons why Embry Riddle was such a great school. The instructors, having had many years of experience improvising, would generously share their wonderful secrets with us as the course progressed. Those secrets could not be found in any textbook.

As the assistant ushered Fadi and Tiny out of the classroom, Fadi barked with outstretched arms pleading innocence, "It was just a joke. The punch came before my punch line."

*

After the assistant left with Fadi and Tiny, the rest of us scrambled to find our wayward, five-eighths socket. I found one under a desk. When I went to my shop table, which was directly beside Bartholomew's, I noticed that the big guy seemed to be just standing there motionless, staring at his open tool box. I thought that maybe, considering his delicate constitution, he had become sort of shell shocked after the dramatic confrontation between Fadi and Tiny.

"Is there a problem, Bartholomew?" I asked tentatively.

"The curse," he uttered almost in a whisper. "I think it's still **cursed.**"

He was, of course, in all seriousness referring to the teaching assistant's humor about crescent wrenches being cursed. I decided to not even try to explain it to him. Observing that the big guy was too large to crawl under the shop tables to hunt for a socket, I gave him my newly retrieved socket and picked up my crescent wrench to continue with the project. Bartholomew's eyes, seemingly with horror, fixed on the crescent wrench in my hand. Did he now think I was cursed because I touched the evil tool?

His eyes then turned to stare silently at the socket I had generously given him. I considered explaining to him that the socket wasn't cursed because I had picked up the evil crescent wrench after handing over to him the innocent socket. But I considered it not worth the effort and continued on with the project.

Both Fadi and Tiny returned to class after lunch and carried on as if nothing unusual had happened. Fadi's swelling eye did not prevent his constant, sardonic grin. I supposed Tiny had escaped an assault charge, that Fadi had been warned about his behavior and that they both were on a sort of school probation. I wondered how often the school had to deal with such matters. I also wondered about our primary instructor.

Chapter 12

The Caddy and the Silver Springs Complex

By the end of Thursday that first week, each of us had completed our airfoil section project. After the rib stitching, we applied the aluminized under-layer of dope, which shrank the fabric tightly to the ribs as the dope dried. The aluminum particles were included to protect the fabric from ultraviolet deterioration.

After that dried, we only applied a clear coat of dope. The assistant instructor had progressively inspected and graded our work at the various stages of the project.

Upon waking early Friday morning, I sensed right away that something weird was going to happen. Maybe it was because the birds and crickets were unusually silent.

By the time class started that morning, I was quite anxious. As the assistant instructor began his lecture on fabric repair procedures, I noticed that the soothing breeze wafting in through the large, open, classroom windows stilled. The internal sentry of my mind knew that something weird – even sinister – was almost upon us. We heard louder than normal footsteps clomping ominously our way in the hall. My back stiffened.

Suddenly, the classroom door burst open. There, filling the doorway, stood the boot salesman, primary instructor, partially crouched, legs spread wide, a joker-like grimace as if a cartoon mask attached to his face. My attention then focused on the large hunting knife grasped menacingly in his hand.

We, the students, naturally froze in our seats. The assistant instructor just stood there expressionless. I wondered if the boot salesman's mind had snapped and he had come for revenge for the complaint. I also entertained the notion that this was to be my personal doomsday. I was only eighteen. What a pity, but I resolved to go down swinging, at least between the whimpers.

Tiny stood up. Yea Tiny, our super hero saviour. But Tiny didn't move. The two of them stood still and stared at each other. At least bark, Tiny.

Suddenly, and shockingly, the crazed boot instructor exploded with a loud burst of explosive laughter – the laughter of the insane. Lifting his knife threateningly high, he launched his body forward toward Tiny. Tiny crouched defensively, his body stiffened. The crazed instructor then suddenly veered toward the side of the room. Running with loud clomps, he skirted around the class toward the shop area.

We all mutely watched as he wildly screamed and plunged his knife several times into all our projects, tearing the fabric and breaking several of the projects' spruce ribs. He then ran off, shrieking as he exited the shop door.

We all turned to observe the relaxed, deadpan face of the assistant instructor. He merely said, dryly, "Make sure you copy down these notes about the repair of doped fabric and spruce rib repair. You will begin the repair of your projects this afternoon."

He casually turned and began writing and sketching on the board. Some will suspect I invented the incident to enhance the dramatic effect of the story. All I can do is invoke the still valid cliché that real life is often stranger than fiction.

So okay, I said to myself after I resumed breathing again after several stunned moments of nerve decompression, this was not to be one of those fine day to die episodes. The incident even had its benefits, one being that we would now bask in the after-glow of a cortisol induced endorphin release.

After taking note of the assistant instructor's calm demeanor, most in the class, as discussed by us later in a more casual setting, recognized that the incident had been intended as a staged humor attempt to lead us into the repair stage of the project.

We supposed the abstracted ethic here was to impress on us that real mechanics must possess the mental toughness to easily take such nonsense in our stride, so to speak. Those who do not possess such toughness should take note that perhaps a more prissy occupation may suit them better and not waste their time any further at the college – either that or they should transfer to one of the more cerebral, academic engineering programs. Sorry, Texas - just kidding.

Real mechanics must possess the ability and self-restraint to calmly deal effectively with customers who sometimes lose control so far as to explode outrageously. As well, mechanics must also be willing and able to sometimes perform filthy tasks, and perform them under excessive conditions.

One initial theory was that both the boot sales episode and this dramatic incident had been staged. However, since our sleuthing revealed that attempted sales of boots in class with a maniac follow-up had never occurred before, and would never occur again, we concluded that the primary instructor did indeed suffer a nervous collapse of sorts, with this intense incident unadvisedly staged to cover up the fact.

*

That Friday, after class, the guys at the hotel split up into apartment mates for shared accommodation at the much cheaper Silver Springs Apartment-Hotel complex directly across from the Eastern Airlines maintenance facilities at Miami International Airport.

My hotel roomie, Texas, teamed up with three fellow engineering students. Tiny's roomie teamed with fellow business pilot students, so Tiny and I teamed together. Most of the other guys were afraid of him, but I had no problem with the big guy. I think he sensed that, and that's why he sought me out.

Then we had to team with two other students for a two bedroom, bunked apartment, the cheapest way to go if we could manage to avoid killing each other over the natural differences of inexperienced young people.

Tiny had come up to my room from the hotel's third floor to ask if I was agreeable to sharing accommodation. It was when we headed out to go down the stairs for dinner that we discovered Bartholomew and Fadi blocking our way, as if they had been waiting for us. Bartholomew gazed down at the floor where his foot, as if absently, traced out small circles. Fadi obliquely studied the ceiling.

Finally, Bartholomew spoke. "We, Fadi and I, are wondering

...humbly wondering... if you gentlemen would be kind enough...kind enough to accept us as...accept us as your roomies at the Silver Springs complex. Fadi can cook, and he just bought a big car."

Tiny and I stood there mutely stunned for a few long moments. Tiny remained speechless, so I assumed it was up to me to respond.

"The apartments are rented by the month," I began. "Let's try it for a month. If it doesn't work out, we'll try a different arrangement."

"Roomies," responded Fadi, looking at us now without the usual wolverine grin. He was on his best behavior for the moment. "Let me take you for a ride in luxury."

We descended the stairs together. Fadi can cook? This should be interesting.

*

I would learn that Fadi was the son of a fairly rich Lebanese in Lebanon merchant. Fadi's father agreed to his son's passionate pleads to attend Embry Riddle for two reasons. One was to stop his son's incessant carping about it all the time. The education would take only half the time of a degree. Universities were useless anyway for the tough realities of the business world. The boy could learn the essentials for his business place in the family on the job after he got the aircraft nonsense out of his system.

Also, the father had been informed that aviation people are usually taught to especially feel empathy for their fellow humanity. They are taught to be very, very careful to make aircraft safe for people both in the air and on the ground. They treat airplanes as if they were carrying their own family and friends, or as if their family and friends were in the houses over which the airplanes would fly.

So Fadi would become a good person who does not just think of himself, although he must naturally preserve sufficient shark in himself to be successful in business.

The most important reason for his father agreeing to send him to far off Miami, though, was for Fadi to very hopefully meet, charm and marry an excessively rich heiress student of the prestigious, all female, Barry College, rumored to be the Catholic version of Sarah Lawrence.

My understanding was that the college was favored by the Catholic rich and famous as a cloistered fortress to send their daughters for safekeeping until their career or marriage. At least that was the usual perception of the parents. The college was run by a convent of Nuns, but the college itself was not a convent. I'll describe it in more detail shortly.

*

So we four continued down the hotel stairs for a wonderful, open-topped Caddy ride to the Silver Springs complex to rent an apartment that very evening. Outside, Fadi tossed the keys to Tiny. It was good that Tiny did not perceive Fadi's mind game that quickly began with Fadi barking out directions to the complex. In doing so, Fadi was directing, controlling the big guy. It was fun to watch, but I didn't hold much hope for our new arrangement.

I knew one corner of Bartholomew's psyche. I would soon learn that other corners of that psyche were interestingly inhabited. That's soon to come yet too.

After we arranged for our accommodation at the Silver Springs complex, Fadi directed our setup shopping. We shared equally in the cost for food and cooking supplies, but bought our bedding separately.

Back at the apartment, Fadi cooked up a tasty Lebanese meal that featured what to we other three was a surprisingly delicious, mysterious mixture of exotic herbs and spices. The spices were minimized to accommodate the standard North American plain palate of overcooked meat, mashed potatoes, a serving of a limp, dead veggie and a desert with lots of sugar to putrefy the meat.

Fadi knew that our new palate was to be educated to the new tastes gradually. It was so delicious that we culinary Philistines didn't miss sugar at all. A diet like that would necessarily shrink Bartholomew's girth. We other three rotated the washing, drying of the dishes and serving as Fadi's assistant cook flunky.

Tiny did not object at all to Fadi's den mother role. Fadi also directed that we perform an apartment cleanup every Sunday, including the once a week washing of our clothing and bedding. Also, every Sunday, we settled our collective account with Fadi as account manager, of course.

*

After dinner we all piled into the Caddy and, with Tiny driving while Fadi barking out directions, we drove over to the airport terminal where, from an observation deck, we enjoyed the fantastic pleasure of just watching propliners like DC-7s, Constellations, DC-3s and Martins start their powerful engines.

No one then imagined that the initial, billowing cloud of engine smoke was contributing to global weather extremes. We just stood there gazing, almost glassy eyed, as we silently responded viscerally to what we each had in common, a deep awe of aviation with its mesmerizing rituals, such as startups.

With the Miami summer early darkness rapidly coming on, we observed below us a cockpit with the crew beginning a pre-start check, another ritual. The multi-coloured panel lights lit up magically.

After observing that for who knows how long – time while in that ethereal state of mind passes totally unnoticed – we left the observation deck, piled back into the Caddy and headed for the airport's south side on Perimeter Road.

We backed onto a treed strip beside the road, allowing a very close-up view of runway 27 where a constant line of airliners accelerated their engines to a body-vibrating, passion arousing, full roar takeoff

Other cars were there too because the strip also served as a lover's lane. When the police drove slowly by with a flashlight shining to ensure respect for the naughty laws, they knew that a car full of guys was an innocent Embry Riddle thing, although, with our being in Miami, not St. Paul, they at least gave it a high probability.

The trees of the Perimeter Road strip are gone now, replaced by the Dolphin Expressway. That does not alter or diminish the memory.

These rituals had to be repeated by us at regular intervals. As with all inducers of euphoria, once is never enough. We returned to the complex, more optimistically bonded now.

Chapter 13

Fast Eddie

Monday morning, we four hopped into the Caddy, with Tiny driving, of course, to begin our second week at the college. As we pulled out onto NW 36th Street from the complex, the Lockheed Electra airliners at the Eastern Airlines maintenance facility loomed large in front of us.

I felt envy for the technicians, out in the fresh air, enjoying the unspeakable privilege of climbing all over those beautiful airplanes.

My inexperienced mind did not speculate as to what those technicians had to endure on very hot, humid days, or when the pressure was on. For example, the flight must go, but needs a part not presently available. A person without conscience and/or without the imagination to realize the perilous possibilities, might rob the part from a plane coming in for maintenance, and then rob the robbed plane to replace the part when the flight comes back later.

That would be naughty because anytime a part is removed, it has to be sent in for re-testing and re-certification before it may be used again. Airline safety depends on procedures like that. I should add that I've never seen that done at a major airline.

I was aware then that those Electras had developed a very serious problem very recently. I'll get to that in the next chapter. I was also aware that the very famous Eddie Rickenbacker was the major owner and Chair of the board with Eastern Airlines. I'll describe Fast Eddie for the rest of this chapter.

*

The reasons why I consider Eddie a noteworthy person are that he particularly represented the American dream, and that he was the primary symbol of the enormous change in aviation which was taking place in that era.

When we consider Eddie's early impoverished youth in America, we naturally wonder how such a person become a major owner of a major airline and the Indianapolis Speedway. As a boy he immigrated to Columbus, Ohio in the very early twentieth century with his German speaking, Swiss parents and siblings.

There, he spoke broken English, was so poor that meals were missed, and he had to stay at home from school, huddling with his siblings for warmth on the coldest winter days because they couldn't buy winter clothes. That did not appear to stunt his growth.

As well, he fought off bullies, and, at fifteen, left school to take on head of the household responsibility as the family's only breadwinner when his father died.

His ascendance to the summit of the American dream was accomplished with nothing more than an extraordinary intelligence, an exceptional intuition, drive and energy, charm, a commanding presence, a six foot two stature, natural leadership, astounding courage, bravery and luck.

Some have the Dream handed to them. Eddie had to earn it, and he was eminently equipped to do so.

*

Columbus, Ohio then in that first decade of the twentieth century, was a great place for innovative technology. The young Eddie went from industrial job to job but learned technical skills as he went. He charmed the techy pros into revealing their techy secrets. The intelligent fellow, still a teenie, invented ways to increase production, which he applied particularly at a car manufacturing job that boded well for his future.

That car manufacturer also built race cars, racing having become a public craze by then. Eddie first became chief mechanic for the company's races, and then became a driver.

He won with an underpowered car because he learned to get through the corners faster than anyone, so the foreign manufacturers offered him more money to race for them. He patriotically refused and won enough races to collect plenty of the tens of thousands of dollars of prize money. He did not blow his money. He invested as a part owner in the fledgling Indianapolis Speedway.

Many race drivers were killed then. Without seat belts, they hurtled through the air at ninety miles an hour after a miscalculation, usually not surviving. Inevitably, even the brilliant Eddie crashed, but he survived it. Eddie would have that kind of luck on other occasions as time went on.

He was already independently wealthy and famous in his mid to late twenties when World War One came. Even so, the American Air Force did not want him because all the selected pilots had to possess a university degree. An incomplete elementary school education didn't cut it. A lot of smart guys bit the dust.

Eddie worked around it by showing how stupidly their airplanes were being maintained. It was then a quick hop to becoming, not only a fighter pilot, but America's chief combat Ace with the included, enormous fame that piled on top of his already acquired racing fame. He was then the most famous person in the world.

He was chosen to command those college grads. Added to his own wealth was that of an older, wealthy wife with most excellent connections. That was not a cynical calculation on Eddie's part. He really did love her, and she him, for the rest of their lives 'till the grave do part.

*

Eddie became that American, number one combat Ace mainly for three reasons. One is that he could fly his open air cockpit fighter higher than anyone else.

His years of shivering with his huddled siblings in the unheated house in Columbus allowed him to endure the pain of the progressively colder and oxygen depleted air at altitude longer than others, especially the coddled college grads. The higher altitude gave him the combat advantage.

Another reason is that he used his intelligence to pick the moment of attack with skill. The concept is called 'intelligent risk'. A person acts to the maximum of aggressiveness without personal destruction. Of course, one needs great skill to accomplish it.

The Red Baron was an Ace, but his method was to fly above the fray until one of the less valiant tried running away. He then swooped down from above like a hawk to snatch the easy prey. He was just doing like all living creatures do in nature. Perhaps some historians may disagree, but it's my researched understanding.

Adding to Eddie's high pain threshold and grasp of intelligent risk, the third reason why he excelled in combat was a combination of aggressive leadership mixed with instant pattern recognition derived from an intense, but broad focus of peripheral vision.

I'll explain, starting with aggressive leadership. Since Eddie could fly higher than anybody and insisted on leading the guppies (college grads) into battle, he took off and flew high before the others, and headed toward engagement while the guppies, with their pink little lungs huffing and puffing, climbed to an altitude they could handle all the while keeping that little dot of their leader in sight.

When they saw the dot suddenly plummet from the sky, they knew that the enemy leader was about to perish and that Eddie would induce the confused others to swarm after him, putting the tails of those confused ones conveniently in the sights of the grateful guppies. The guppies were fiercely loyal to their fearless leader who was incapable of letting his guppies fend for themselves with himself circling safely above the fray.

As for the instant pattern recognition derived from the broad focus of peripheral vision, I'll refer to sports figures and airport control towers.

The greatest NHL hockey scorer of all time was Wayne Gretzky. Wayne did not have to look at the puck when on his stick. Rather, his eyes were in constant peripheral scan. We could compare it to the situation in an airport control tower where a controller simply scans the radar screen to know exactly where everyone is, where they're headed and at what speed. Gretzky saw, in a glance, where everyone was, what they were doing and what they were about to do. His scoring stats were crazy impossible. The legendary football quarterbacks played similarly.

I should add that the planes Eddie and the guppies flew were the obsolete, underpowered castoffs of the French and British forces, which were certainly very inferior to the German planes.

We could compare it to the underpowered race cars that Eddie drove. He figured out a way to win, especially with his cornering ability.

One combat incident especially illustrates Eddie's character. On one of his plunges in his inferior plane from high altitude to engage the enemy, the fabric on the top wing of his biplane ripped off. That would probably have ended the earthly experience of any of the guppies or anyone else.

Only Eddie would have realized that he needed to accelerate in his downward plunge to get enough increased airspeed to compensate for the reduced lift of the remaining functional wing. He slammed the throttle forward, noted with amusement the enemy soldiers on the ground shooting at him, yawned, and lived to fly another day. So okay, maybe he didn't yawn.

All that is why Eddie Rickenbacker was America's best combat Ace. There was more to come.

*

A major factor in Eddie's successes, whether commanding a fighter squadron or in business was his ability to handle people – to get them to perform or to invest in his usually successful business enterprises, or whatever else.

An example of whatever else occurred during World War II when the American President (Truman) became alarmed at the bellicose behaviour of General Douglas McArthur. To persuade McArthur to tone things down, the President considered the no-nonsense, tougher than nails Eddie as the only person in America with the overpowering personal presence to prevail in a face-to-face confrontation with the General who had an overwhelming personal presence of his own.

While riding in a specially requisitioned bomber for the over Pacific Ocean jaunt to meet with the wayward General, Eddie discovered that the navigator's equipment had failed. They flew dead reckoning, hoping to find that small, very isolated island where they could refuel to complete the flight.

The island refused to appear, so they flew around to try finding it. Finally almost out of fuel, they emergency landed on the water and lashed three rafts together for the eight men. They did not have time to load any food or fresh water on the rafts before the plane sank.

One of the men would drink sea water, go crazy and die. The others captured rain water (very sparse), and chewed on any raw birds or fish that ventured too near them. That allowed them to barely survive at a subsistence level, if they had the guts to accept the pain of it.

As the days, scorching sun, hunger and thirst went on, Eddie was naturally the one to control their courage to endure. The captain was not the real captain. It was once again as if Eddie was the commander of guppies.

However, the original guppies admired their commander. As two weeks went by, this situation went to a different place that required different leadership. Any admiration would be delayed.

When anyone said anything at all that sounded negative, Eddie would bark, "What was that?! What was that?!"

followed by an insult to their manhood. They all came to hate him because of his nastiness. To Eddie, hate was good. Hate caused them to focus on him and not their misery. Hate was essential in that hostile predicament.

Even that would probably not have saved them if Eddie's wife had not intervened. Discovering that the rescue search was called off after two weeks, she went directly to the President and threatened to rip his head off if he did not extend the search for at least one more week. Eddie and his wife were well matched.

They were found during the third week. Although the other men then disliked Eddie intensely after the tortuous experience, they all agreed that they probably would not have survived without him. After a face to face meeting with Eddie, General McArthur proved a total pussycat and toned down the bellicosity.

*

In addition to Eddie's achieving the American Dream, I earlier indicated that he served as the primary symbol of an enormous change in aviation. Slightly before my arrival in Miami, Eddy was removed from his position as president of Eastern Airlines. He became chairman of the board as a sop, but he no longer had control of the daily running of the business, such as deciding which airliners to buy.

Eddie's fantastic business intuition had finally failed him. He was a mortal human being after all. He was convinced that jet airliners would not catch on with the public. Bogus. Although he was okay with the relatively small, short hop turboprops such as the airline's Electras because of their short field takeoff and landing capability with relatively high payloads for that performance. Turboprops are, of course, jet engines designed with all the thrust coming from a propeller. They were relatively economical engines as well, especially compared to the fuel hungry turbojets.

Back then, aviation fuel was seen as lasting forever with a never ending supply. A carbon footprint was the residue an animal's paws left on the ground after the animal had walked through an area recently subjected to a forest fire. Efficient turbofan engines were then yet to come.

It was that reluctance to invest in the new jetliners that Eddie Rickenbacker served as the primary symbol of that enormous shift in the airline industry's conversion from piston powered propliners to jetliners.

Now I'll explain why those Electras I saw lined up at the Eastern Airline's maintenance facility were also of special note.

Chapter 14

ELECTRA

Most aviation enthusiasts know that Lockheed produced two types of Electras. The first one, the Model 10, coming out in 1930s, was an early airliner that looked similar to a Douglas DC3. It was powered by two radial, piston powered engines and seated 10 to 12 passengers. One of those Electras achieved fame as the one in which Amelia Earhart disappeared in a Model 10E.

The Lockheed Electras I saw at the Eastern Airlines Miami maintenance facility long ago had the designation L-188. Introduced to service in the later 1950s, it was a 70 to 90 seater airliner powered by four (Alison) turboprop engines of 3750 eshp mounted on relatively stubby wings.

Its short field performance was/is exceptional, and it was/is known as a pilot's airplane because of its maneuverability.

*

The word maneuverability prompts a digression to this separate section. If you already know how an aircraft's aerodynamic centers (centers of movement or rotation as in pitch, roll and yaw) and structural centers (centers of gravity of fuselage and each wing) relate to maneuverability, efficiency and sudden death, then skip to the next section for a continuation of the L-188 story.

However, if you're a beginner pilot thinking of going directly from a training aircraft to a performance airplane, then this section is mandatory reading.

The great maneuverability of the L-188 Electra was due the longitudinal and lateral centers of gravity being located so close to the centers of movement or rotation. Therefore, little resistance is experienced in roll, pitch and yaw. It follows that control responses are very quick.

To illustrate, try holding a mop straight out from the handle's far end. Much effort is required. It's called a long moment arm. Move the mop up and down. The movement is slow.

Now hold the handle close to the mop end and move it. With this shorter moment arm, the effort is less and the movement is quick.

The reduced resistance to movement also translates to fuel efficiency. For example, with the aircraft longitudinal center of gravity located so close to the center of lift, less downward force at the tail is needed to keep the plane flying level.

The inherent instability of close centers is why modern fighters need computers for control. Otherwise, the human brain cannot react quickly enough to prevent the increase in flight oscillation that leads to catastrophe.

A smaller performance plane, like a Beech Bonanza, brings its centers close enough that an 'experienced' pilot can handle the controls without computer assist. Sufficiently experienced pilots can do that because their gradually progressive training over time has gradually grown and marshalled enough brain neurons and synapses to handle the quick input responses needed.

It's why race car drivers need that long, progressive development. It's why relatively inexperienced pilots need to begin flying with flying pigs – very slow, very stable airplanes.

A flying pig, like a Cessna 172 and the like, have their centers of gravity and rotation safely stretched out. The controls feel sluggish to a pro, but quick to a rookie. With such stable aircraft, if a control input is made and then the control yoke released, the plane tries to return to its original direction and automatically dampen flight oscillations.

A performance airplane, its centers of rotation and gravity closer together with short moment arms, tend to stay in the input direction pointed. The reduced resistance also increases its rotation and flight velocity much quicker.

It's the reason why a performance airplane like the Beech Bonanza is called a doctor and lawyer killer. 'Some' of those professionals have the net income to go straight from a training Cessna 172 to their own seductive, gleaming Bonanza with the gratifying rumble of a much larger engine.

Some of those professionals, and even some impecunious types who have sold themselves into financial slavery for the rest of their existence, have not taken the intermediate steps to train their brains adequately to handle the vastly quickened control maneuvering and aircraft velocity increase.

The problem is very evident in soaring (gliding). Most begin in a flying pig with safe, thick wings that have baseball bats for leading edges and long moment arms. Some who proceed from that to a performance type find that the increase in velocity is so quick when maneuvering that their underdeveloped brains cannot react quickly enough to avoid a do not exceed speed. The wings rip off.

I hope that information will save a life or two, as well as the lives of any innocent ones going along for the ride. Let's return to the L-188.

*

As the Caddy rolled onto Miami's NW36th Street to begin our second week of classes at the college, there were some things I did not know about Eastern Airlines then, and one glaring, life and death, matter that I, and every aviation person then, did know.

I did not know that the L-188 Electra received its type certificate only two years before my arrival at the college, that the first of forty Electras was delivered to Eastern two months after the type certification and that the final delivery would be a couple of months after my arrival at the college, that a recent development then was causing Eddie Rickenbacker's Eastern Airlines, Lockheed, and the other L-188 Electra owners much stress.

About ten months before my arrival at the college, witnesses on the ground near Buffalo, Texas, observed a turboprop, which had been cruising straight and level at altitude with no sign of a problem, suddenly explode in a fireball and disintegrate.

The Braniff International Airways airliner, an L-188 Electra, originating the flight in Houston, had been bound for New York, with intended stops at Dallas and Washington.

With there being no cockpit voice recorder, or air data recorder, and a disintegrated Electra, it would take the CAB (The Civil Aeronautics Board. The NTSB was not created until 1967) time to figure out what had happened. Cockpit voice recorders had been developed by then, but pilots wanted their conversations to be private. It would take a while yet to work out that issue.

The CAB quickly determined that with the flight having less than half the passenger capacity, and the plane being 16,000 pounds under maximum weight, structural loading was not a factor.

With the plane having been observed flying normally straight and level at cruising speed, aerodynamic loading was also not a factor.

The weather had been good and clear with minimal turbulence with visibility ten to fifteen miles, so a freak storm did not tear it apart. No periodic inspections had been missed.

The explosion and fireball mainly meant one of two things. Either a catastrophic structural failure had caused the fuel tanks to explode, or one of the varieties of psychopaths had arranged for the explosion. Life insurance policies would be explored.

In that era, very few seriously speculated about the possibility of an errant missile or a zap from an alien pod.

One must not jump to conclusions. Negative consequences often occur after such jumps. Sound conclusions must rest on solid evidence. Time went on as the CAB continued investigating.

*

Then, about five months after the Braniff crash, or five months before my arrival at the college, and to everyone's extreme alarm, a Northwest Orient Airlines L-188 Electra came apart during cruise near Tell City, Indiana beside the Ohio River. Again, a wing had separated and was found five miles from the fuselage, or at least what they could find of the fuselage in that crater.

Witnesses on the ground said that the right wing separated, a mirror image of the Braniff crash in which it was the left wing that had separated.

As with the Braniff crash, various causes were considered. This time some turbulence from storm cells had been in the vicinity, and a National Airlines airliner did suffer an onboard dynamite explosion only a couple of months previously (*Look for the strange case of Julian Frank below which reveals some peculiar facts and speculations about that crash).

Those possible causes would be investigated, but it seemed most probable that the cause was another wing separation that knocked the empennage (tail) off with further disintegration in the plunge.

Back then, economic considerations trumped crash probability stats. Instead of grounding the entire fleet, an AD was issued. Aviation people know that a government issued AD is a mandatory Airworthiness Directive to correct an immediate danger.

The first, temporary measure, AD was issued three days after this Northwest crash. The maximum structural cruise speed, the Vno was reduced from 324 knots CAS to 275 knots. Five days after that, another emergency AD was issued which reduced the Vno further to 225 knots and the never exceed speed Vne to 245 knots.

Also, the pilots were to not use autopilot and were to closely observe the torque meters readings. On turboprops, the torque meters measure how hard the engine/prop combination is working. The pilots were to feather the propeller of the engine if the meter showed a torque of either zero or full scale. That must have been a very tense time for subsequent to the crashes Electra pilots, of course.

For those interested, I'll explain terms like CAS, torque and feather at the end of the chapter.

*

By then, aircraft manufacturers thought they had solved the problem of airliner's coming apart in the air. Everyone in the aviation world then remembered very well how the world's first jet engine airliner, the de Havilland Comet, disintegrated during cruise in 1954 – a BOAC Comet, followed only weeks later by a South African Airways Comet also disintegrating during cruise.

Investigation revealed that the Comet's design safe loading was too low, the skins were too thin, the squarish windows concentrated too much stress at the corners, the window supports should have been glued on as designed, not riveted, and that the punch rivet construction technique created tiny stress cracks that would soon not be so tiny.

Also, the structural fail safe design was subsequently developed in which a partial failure would remain isolated and not tear apart the whole structure.

Other airliner manufacturers subsequently learned very well from the tragic set of accidents and included the improvements in their own designs. They observed that the same tragedy would most likely have happened to one of their products if it had not happened with the Comet first.

Comet sales naturally plummeted, but de Havilland included the improvements, rounded windows and all, in the Comet's future versions, including the military Nimrod version.

One design feature of even the improved future Comet versions that the other airliner manufacturers did not want in their planes was the engines being embedded in the wings. It was thought that hanging the engines in pods beneath the wings, even if it added to the inefficiency of extra drag, would be safer in the event of a fire.

*

So what were the ultimate findings of the L-188 Electra investigations? The problem had its undetected beginning right after that four engine turboprop was initially put into active service.

The passengers complained of an annoying level of vibration and noise in the cabin. Lockheed's solution, after further research, was to cant the engines three degrees upward. Testing the new configuration did not reveal the hidden problem. The subsequent passengers found the remaining vibration and noise levels reasonable (Aviation enthusiasts like me love all the vibration and glorious noise we can get, but, of course, I know most people do not).

One finding of both crash investigations was that the engine mounts had stretched. Lockheed then found that by testing the engine/propeller parameters beyond what was the accepted norm for the industry, a condition they called "whirl mode" was induced.

Under certain conditions the engine/prop combination began to wobble. The wobble induced a wing flutter. The flutter, combined with the wings' natural frequency, harmonically coupled to the flutter frequency, stressed the engine mounts beyond their design capacity. The mounts then weakened, producing more wobble. The wing attachments were then stressed beyond their capacity. The wing ripped off.

The concept, minus harmonic coupling, can be illustrated with a toy called a top. In this digital age of fewer mechanical toys, I'll explain that a top is round in one view but is an inverted triangle in another view, so comes to a point at the bottom end. Pumping the handle at the top causes the toy to spin, after which, a magical force causes the toy to remain standing on its point all on its own without having to hold it upright.

If we then slightly nudge the top on the side, a resultant force called precession will cause it to veer slightly at an angle ninety degrees to the direction of the nudge (Helicopter people know all about precession in a whirling system). The toy remains spinning on its point, but it now wobbles.

Lockheed engineered the fix, which included strengthening the engine mounts among other improvements including the adding of a fifth engine mount. There were no further such problems with the Electra. Sales of the airliner naturally fell off, but pilots loved the plane. It handled beautifully, and it could get in and out of fields that comparable aircraft could not.

The American military purchased a military version called the P-3 Orion which is used mainly for maritime patrol and research duties. It's one of the patrol aircraft used for flying deliberately into hurricanes to determine the storm's path so resources can be effectively deployed to reduce the damage. It was the aircraft that flew into the most powerful hurricane to threaten the American east coast. The forces on the plane were horrendous. The pilots magnificently flew them out of what seemed for a while like certain disaster. The wings handled the extreme loads magnificently. Lockheed's original fix and further version tweaking worked just fine.

*

So what happened to Eastern Airlines? Eddie Rickenbacker had been ousted as CEO only two days after the Braniff Electra crash, the first one to lose a wing. I can picture Eddie pointing his bony finger at the Eastern Board members and chiding, "I told you! I told you! First it was the Comet, and now it's the Electra. Jetliners, including these jets pretending to be propliners are a disaster. The public won't accept them. We're going to lose our investment."

He would be referring, of course, to Eastern's just completed purchase of forty Electras. No doubt he had been very reluctantly persuaded by the Board members and Lockheed to buy those planes.

The Electra made, on paper, economic sense with its efficient operation and ability to get in and out of small fields. Commercial people would appreciate avoiding the hassles and inefficiencies inherent in the large airports. As well, it was Eastern that pioneered hourly commuter flights to NYC, Washington DC and Boston.

It was Eddie who had proudly guided Eastern to being a profitable airline that became the fourth largest among the major carriers. Eastern was apparently the only one that did not require government subsidies to remain solvent. His being ousted as CEO must have been an enormous blow to him.

The new CEO, Malcolm MacIntyre, was also representative of a major transition in the American airline industry, just like with the transition to jetliners. Malcolm was considered a brilliant lawyer but was airliner inexperienced. He represented the new professional class of airline executive who could better navigate the increasingly complex financial world, but would have to rely on his aviation type underlings for those particulars.

That would erode certain accepted ethics, understandings and ways of doing things in the industry. For a parallel understanding of how that way business was changing generally in America and the world, read about how the Morgan financial institutions changed with their evolving ethics – hostile takeovers and more.

Eddie would hang on as Chair of the Board for three more years until he was seventy-three. During that time, beginning in 1960, he saw the introduction of the company's jetliners, the first being the DC-8-21. Eddie lived on into his mid-eighties, quite happy with his beloved older wife.

As we know, and as the great Eddie Rickenbacker came to humbly accept, the public did indeed accept jetliners. Flying above the weather for increased safety and getting non-stop to the destination quicker with greater efficiency appealed to all. Propliners barely held on for a while, but were soon doomed with the major operators.

The public also accepted the L-188 Electra itself. We seemed to have confidence that Lockheed would quickly fix the problem. They did.

*

Eastern carried on and even grew for a while in the next three decades before its financial demise. Those L-188 Electras would remain in service for another seventeen years after the crashes. Eastern became the official airline of Disney World in 1971. In 1961 they entered the turbojet airliner market with a sizeable investment in fifteen B720s, and did it when Eddie R. was still Chair of the Board. Maybe that was what brought on the hurricane that year. In the 1980s they were the first customer for the B757.

The first sign of financial trouble for Eastern followed the Airline Deregulation Act in 1978. That Act resulted in a slug it out free-for-all in the American aviation industry. It was like a Wild West bar brawl followed by a shootout.

It was like what happened with the 19th century American railroad industry when everyone and their dog tried to set up their own airline. Fares plummeted, which benefitted the public financially. Operations like People's Express, for example, offered very low fares by offering 'no frills' flying.

To remain solvent, those types of airlines paid lower wages, and scrimped on everything scrimpable. That meant they hired relatively inexperienced personnel to both fly and maintain the aircraft.

The larger airlines, like Eastern, advised the public that the higher fares would mean their pilots and technicians were more experienced. More experience translates to more safety for the passengers. Also, frills make for a more enjoyable ride.

Having become financially overextended by the time of deregulation, Eastern had to scrimp where they could also. That, for instance, led to a labor dispute, followed by a layoff of four thousand personnel.

Like a weary boxer losing the fight, Eastern stumbled on as best they could, hoping desperately to land one great punch for an unexpected victory. For instance, they sold off their more Western assets to concentrate the remaining resources on their original East coast roots. It was too late. Their last gasp came in 1991. Mercifully, Eddie did not live to see it.

*

One theme of this memoir is the transition of airliners from piston engine propliners to jetliners. So when did the major airlines retire their piston fleets?

There was a period of overlap. Turbojet jetliners like the DC-8 and Boeing 707 were introduced in the late 1950s. The piston engine propliners were phased out in the mid-sixties to the late 1970s.

*

I'll complete this chapter with the related, peculiar case of Julian Frank. A few weeks before the second Electra crash with the whirl mode problem, a National Airlines Boeing 707 was scheduled for a flight from NYC to Miami.

However, one of the 707's windshields became cracked. On a first serve first come basis, 76 of the early passengers boarded an L-188 Electra. The remaining 28 latecomers boarded a slower DC-6B which also was inconveniently scheduled for a stop at Wilmington, South Carolina.

Julian had a lot on his mind. Perhaps he was one of the later ones because he was hesitant about making the flight. For one thing, he had an extreme fear of flying to the extent of induced panic attacks.

For another, the Manhattan District Attorney's office had accused him of misappropriating about a million dollars in several charity scams. He had that very day bought nine hundred thousand dollars in life insurance. One could say that the outlook for this flight did not bode well.

The L-188 Electra flew safely to Miami. I'm sure the nervous pilots paid a lot of attention to the engines' torque meters and flew at the reduced speed.

The DC-6B stopped at Wilmington, NC for fuel and extra doomed passengers, and then took off. It was near the town of Bolivia, NC, south of Wilmington that the explosion occurred. Julian Frank, or at least his torso without legs, was the only passenger ejected. He was found sixteen miles from the final crash site.

The plane had made a wide right turn before crashing, perhaps in an attempt to return to Wilmington for an emergency landing. A witness to the accident reported hearing an engine cutting in and out. I suppose the explosion ruptured the fuel supply to one side, and maybe debris damaged a prop.

Some of Julian's fingers were found with the airplane at the crash site. His hands and body had the residue of dynamite and dry cell batteries. Small electrical wiring consistent with an explosive device were embedded in his body. Suspicion of Julian's motives for boarding the flight loomed largely.

What could not be absolutely determined was whether the device had been on Julian's lap or under his seat. Maybe he had placed his hands under his butt for warmth or some kind of infantile soothing prompted by his fear of flying.

If he was indeed the culprit, we might ask why he waited all that time to get to and leave Wilmington way down in North Carolina. Maybe he needed time to work up the required nerve, but it meant that the new passengers picked up at Wilmington would be included as victims, which appears dastardly evil, although, it's possible he was totally insane that day – after sanely buying the extra life insurance that very day, of course.

Maybe one of the new passengers at Wilmington placed the device under Julian's seat. If so, Julian would suffer the incriminating evidence, and the real culprit would escape suspicion with benefits going to his or her family.

Ultimately, the legalities would remain unresolved. Perhaps Julian's large insurance policy bought that same day was a merely normal purchase to benefit his family in the event of a crash or some other relatively innocent reason like treacherous weather or a parts flaw or a maintenance mistake or pilot error.

The dynamite evidence ruled out the immediate considerations like an errant military rocket strike or a mid-air collision or an engine fire causing shrapnel followed by explosive decompression or a prop blade separation causing the same thing or a fuel or oxygen explosion or metal fatigue.

Note: The Airline Pilots' Association had complained that routine flight proficiency testing for pilots, conducted every six months, involved violent recovery maneuvers that put unnecessary stress on the aircraft, hence leading to metal fatigue.

They even speculated that Julian's extreme fear of flying may have induced in him such a severe panic reaction that he may have pounded the window, weakening it, so leading to explosive decompression. All the above were considerations before the dynamite and device evidence.

I don't know what happened with Julian's insurance policy, but the Manhattan District Attorney's office referred the case to the FBI, who then referred it to the Justice Department where it remained unresolved.

As for National Airlines, it had been Eastern's main competition, even having its headquarters also in Miami in the same Fritz Hotel building as Embry Riddle.

Its slogan had been 'Airline of the stars'. That was compared to Eastern's being the official airline of Disney. Fly National and one might see and be in the magnificent presence of a dazzling movie star. Fly Eastern and one might see Daffy Duck, with an occasional glimpse of Goofy.

Eastern got the last laugh. It remained solvent eleven years longer than National which was swallowed by Pan Am in 1980. In turn, Pan Am was swallowed by Delta.

It was the natural law of nature. A fish is devoured by bigger fish that is, in turn, devoured by an even bigger fish. Delta is still prowling the waters.

Chapter 15

Barry College

Barry College is now a co-ed university, with its main campus still in Miami, but back then, it was where mostly well off and/or influential Catholics naively sent their daughters to be cloistered in a female only college, protected from impecunious, dreadful, predatory young men on the hunt for the pecunious, or, sometimes, even just for the fun of the hunt. The rumor was that such young men must first get past a phalanx of ever watchful Nuns to achieve their aims.

The fault in that cloistering concept was that they located the college in a large urban area like Miami that had an all-male college (back then) like Embry Riddle that possessed very many impecunious, socially challenged, predatory young males.

Certainly the Barry College planners must have known the basic, unbreakable, and powerful laws of the universe. For instance, the vast majority of our humanity knows that the positive and negative poles of a magnet attract. And that if those opposite poles each have fine wire wound around them, an exciting electrical charge may cause an attraction that can be immensely potent.

Those opposite forces send out powerful emanations unseen by the human eye. The two Miami colleges, Barry and Embry Riddle, being of opposite gender polarity then, and being substantially excitation wound, vibrated in their hormonal intensity. In the quiet, early moments before sunrise, a person in the proximity of those colleges would hear a distinctive hum.

*

An announcement of a dance to be held at Barry College came early in September. The U of Miami students naturally received an invitation, especially those in the professional programs. It was natural that the Embry Riddle engineering and professional business program students would be invited, but it was a bit of a surprise to see an invitation prominently displayed in the warthog section, resolutely placed there by an authenticated Barry College student.

The common warthog opinion was that a substantial percentage of the young women at the all-female Barry College simply wanted an experience with a young male, especially with one possessing a reasonable amount of wart-hogness. To quote a later pop song, "Girls just wanna have fu-un.." What really mattered was the pheromones of the moment.

Charlie was an ancient former Korean War fighter pilot classmate who was taking the course for the pure love of aviation. Emanating the wisdom of the ages, he gathered us naïve nubiles together to address us on the topic of the Barry college students. When Charlie spoke, we wide-eyed babes in the ways of this world, listened. The scar across his forehead from a crash was a proof badge of his sage infallibility.

*

With Charlie seated on a chair, we gathered in front of him on the shop floor like kids in a kindergarten class. He first spread out his arms as if to gather the wisdom of the universe, and then spoke.

"With regard to the Barry College dance this Saturday evening, I feel I must clue you naïve pups into the reality of the situation.

Some of you may feel that you know about realities already, but I assure you that this is not a normal situation. Some of you are about to embark on a path that could, maybe could, change you unalterably.

Don't for a second think that those young women are like the ones you knew back home, even if you're Catholic.

Religion doesn't matter here. They're just looking for fun. They're not looking for husbands – at least not with you, even if you're Catholic. However, I'll mention that there's a hidden complication with that to which I'll return in a bit.

Those young women – you must never ever refer to them as girls – are future movers and shakers. Some of you, whether you come from humble circumstances or not, may seek this path that you think will take you to riches. You should know that some Barry College students also come from humble economic circumstances. Their loving parents have sacrificed very much for what they perceive as the future happiness and well-being of their dearly loved child.

A probability is that such parents will not be amused if their daughter falls in love with you. I suggest it's only fair that the infatuated both of you sort this out early before commitment – at least I hope, for your own sakes, that you're truly infatuated with each other – or life will become dreadful indeed.

I'll add that genius comes from anywhere there are humans, regardless of social circumstances. The young woman of impecunious means initially may very well become very pecunious indeed due to her very own brilliance.

Another problem is the probable intellectual mismatch. Whether the young woman is financially blessed or not, she will find herself, after the initial insanity of infatuation, quite possibly connected with a person not interested in literature, history, current world events, and so intellectually on.

Be aware of this. The solution, of course, is for blockhead you to let her introduce and guide you to a very mentally stimulating future that you might not have enjoyed otherwise."

One of the economically humble, blockhead students raised his hand to ask a question. It was obvious to all the gathered that, although Charlie was one of their fellow students, his age and jet fighter experience unquestionably qualified him to speak with authority on any subject that could possibly exist in this world.

"Sir," the innocent eyed student began, "do any of the other colleges under the U of Miami umbrella also strut?"

"Not one."

"Who's in the Barry frontal line?"

"They're the ones who just want to have fun, no matter what. They usually recognize that strutters are not to be taken seriously beyond one night."

"Who are in the flanks?"

"Those in the flanks are often less street smart but usually more intelligent in a strategic sense. They close in on the unprotected sides just behind the initial line of strutters where there's an assortment of college types such as those in medicine, law and so on. They like to have fun too, but are more judicious."

"Is the rumor true that there's a Queen Bee?"

"It's true. The Queen Bee is off to one side initially behind their left flank before that flank's advance. The Queen Bee is, as you might expect, prime in her beauty, but must also be highly intelligent and personable. There's always a Nun standing beside her as a protector. Only the most qualified prospect – never as yet a strutter – may even be allowed to speak to the Queen Bee. I'll end this discussion with my best wishes for your adventures."

Fadi spoke up. "Sir, if we manage to mate with the Queen Bee, will we die just like a male drone bee does when he mates with a real Queen Bee?"

"Yes, but just like that male drone bee you mentioned, you won't die until after you've seen your manhood ripped off."

Charlie stood and raised his arms with palms uplifted, as if to utter a profound, almost spiritual, pronouncement, "Let the games begin."

*

`The evening of the Barry College dance, I knew Fadi would be going with at least me, but I was doubtful about Bartholomew, and I didn't know if Tiny would be going either.

With Bartholomew being an evangelical protestant, and Barry College being almost convent Catholic, I thought he might opt to attend one of his weekend church functions instead.

As for Tiny, he was seriously dedicated to a fellow Lutheran girlfriend back in his hometown. Her very attractive photo sat prominently on his desk so that she would be the first thing he focused on after awakening in the morning. I can personally verify that he had been faithful to her.

When Tiny climbed into his suit to go, I was a bit surprised. I guessed he just wanted some innocent fun for the evening. When Bartholomew put on his suit, I assumed he was going to attend one of those church functions that evening.

It was very surprising to me, almost a shock, when he declared his intention to accompany us to the Barry dance. He no doubt noticed that the eyes of both Tiny and me were wider than usual, but he offered no explanation.

Fadi wore the amused smirk of one who possessed the deep secrets of the universe not perceived by ordinary mortals.

The four of us climbed into the Caddy and set off for the dance. At the college's big hall, the Embry Riddle guys were easily identified by their strut swagger which was exaggerated from their usual gate due to the obvious context and Charlie's instruction.

The arrangement of the Barry students was as Charlie described. A center line was flanked by slightly recessed flanks. The flanks would presumably swoop in at the appropriate time.

The Barry students were not leaving the occasion to random chance like at an ordinary dance. Randomness was certainly an element, but it was well contained within the boundaries of that overall strategy.

Light, popular music was playing when we arrived. The tactical arrangement, as I understand in retrospect, was for that pop music to continue for a while to make mixing easier. That would be followed by a short period of ballroom dancing music for the technically adept, such as with the tango that needs finesse to avoid embarrassment, to express impressive moves.

The purpose was also to identify to all the ones who possessed a certain degree of sophistication appropriate for the era. Ballroom style participation was, but not always, an indicator of their elevated social status.

Implicit was that such status should be verified by intelligent appraisal. The Barry students knew very well that sharks were prevalent in such waters.

The ballroom dancing was to be followed by a period of up-tempo tunes that would progress to a heated frenzy. The frenzy would then be gradually quenched with diminishing tempos and sound levels to, finally, the soft and sensual.

We all naturally gazed at the far end of the right flank to see the Queen Bee. As expected, she was indeed very beautiful, a sharply defined, voluptuous beauty. No doubt she also possessed the exceptional, intelligent, humble personal qualities that induced her fellow students to assent to her predominance.

The Queen Bee's Nun guardian was obvious to all, but in a subtle way. She did not wear a Habit. Her attire was more that of a business person with a grey skirt reaching to just below the knees and a quality, white blouse adorned with a silver watch attached above her left breast.

Her auburn hair was tastefully swept back, revealing the pleasant face of a woman about, perhaps, thirty years of age who could even be imagined to be attractive given certain touches of conservative makeup.

With the light, pop music of the initial phase playing, Fadi, Tiny and I selected partners to dance, but observed the protocol of that phase of selecting a different partner for each tune. Any serious intent could be indulged later as the evening progressed. Tiny did not so indulge initially. He really did intend to remain faithful to his love back home. When he did join in, it was with innocent intent.

Fadi's dance partners all quickly identified him as a shark, and so they were the first to take their courteous but immediate leave of him after each dance.

As we expected, Bartholomew did not participate in the dancing, at least at first. However, curiously, I did notice that his eyes seem fixed unalterably in the direction of the Queen Bee.

Please understand that I'm describing what soon happened next because this is a memoir, not a novel. It's a cliché that real life is stranger than fiction, but it often really is, although, I must add that the strangeness in this case was different than what you're probably thinking.

The ballroom dancing phase began with a Cha Cha. That was when the totally unexpected happened. Neither Tiny nor I had been trained in the formal steps of ballroom dancing. We stepped to the sideline.

Fadi, who had been well trained in ballroom dancing back in his homeland, took a few steps toward selecting a certain person for the waltz, but was immediately distracted along with Tiny and I when Bartholomew suddenly headed across the dance floor in the Queen Bee's direction.

Had Bartholomew been trained in ballroom dancing? Did evangelical Protestants engage in ballroom dancing as an acceptable form of recreation then? My mind began cringing at how Bartholomew might react to being rejected by the Queen Bee. Would he suddenly fall to the floor in psychological agony as he had done when he discovered that his socket had been stolen?

My mind, grappling for sanity, decided that the big guy must certainly be heading toward a person near the Queen Bee whom he thought would reasonably accept his offer to dance.

Fadi, his eyes also wide, put up his hand to hopefully signal to his originally selected dance partner that something very important had suddenly come up and that their dance would have to be temporarily delayed. But her face emanated betrayal.

After then raising two palms forward as a sign of deep regret, Fadi immediately changed direction to follow Bartholomew.

Tiny and I stood transfixed as the big boy, very big boy, came closer to his quarry. He seemed to actually be headed directly toward the Queen Bee. Then I definitely did cringe. The Nun, with her eyes growing wider, seemed progressively more agitated as the big guy approached.

Fadi, quickly closing in on his suddenly renegade roomy, reached for his arm, but it was too late. Then, to our surprise and relief, Bartholomew directed his attention and conversation exclusively toward the Nun.

So okay, I guessed he wanted to discuss some kind of religious topic with the Nun. My cringe quotient diminished, but some cringe did remain. What now? Needless to say, the majority of eyes in that room were directed toward them. The Nun relaxed her vigilance and seemed to enjoy what Bartholomew had to say.

When a tango began, we were very surprised, but somehow not shocked when Fadi escorted the Queen Bee out onto the dance floor. Bartholomew stayed talking with the Nun.

Fadi and the Queen Bee displayed much tasteful skill with the tango, in my amateur opinion. The Queen Bee took her immediate leave of Fadi when the tune ended.

A waltz began. At a slow 28 bars per minute and a ¾ cadence, it was the easiest of the dances. I would say that I was quite close to being shocked when Bartholomew escorted the Nun onto the dance floor. Bartholomew maneuvered his way through the dance quite adeptly. I would later learn that Fadi had given his big roomy ballroom dancing lessons when back at the hotel on Flagler Street. That explained Fadi's shouting at him, "Yip, yip, yip," and the crashing against the wall when the big guy stumbled.

When the waltz ended, the Nun and Bartholomew took their mutual leave of each other. I breathed easier. Bartholomew then spent the rest of the evening just standing and observing the dancers. I did not again see him turn his eyes toward the Nun.

Fadi again approached his originally intended for the next ballroom dance. She tried to play the aggrieved role with a first refusal, but Fadi, with his suave gestures and immense charm, overcame the refusal deftly. He even had her smiling within the minute. They stayed a couple for the rest of the evening. I would later find out that she came from the same country as Fadi, and that her family had a thriving merchant business. Two families could very well achieve bliss.

I'll have more to say about that evening at Barry College later. That part of the story isn't over. Stay tuned.

Chapter 16

Showdown

The week before that first Barry College dance, the macho showdown around the pool of our Silver Springs complex happened.

The four of us were resting on poolside reclining lounge chairs after a refreshing dip when four bodybuilders approached us. They had started a new workout gym just up the street and had just moved into the complex.

They stood there directly in front of us grinning for a long moment as if they owned the space and that our only chance of sharing that space with them thereafter would be to accept a lower level macho rating than them. They were supposed to be alpha and we were to be beta.

Lounging beta Fadi pretended sleep behind his sunglasses. Bartholomew and Tiny also stayed unreactive lounging.

I was the only reactive one for the moment. Sensing possible danger as the muscle guys approached, I sat up with my feet flat on the floor on each side of my lounge chair so that my center of body gravity was slightly forward. That was so I could launch quickly if I had to, but only for self-protection, of course.

A couple of years previously, I had taken Jujitsu training at the YMCA in the Sault. That discipline was meant only for self-defence, the type of self-defence that emphasized the avoidance of my personal, physical harm only.

Disciplines such as Judo and Karate then added the removal of threat by disabling attack. Our instructor stressed the importance of avoiding conflict if possible, that it wasn't a weakness, that indeed it was a strength to resolve a conflict without violence.

An attacker would soon regard the attrition strategy with respect, not look for vengeance due to an injury, and never try attacking that person again who possessed the superior defensive ability that could make him look stupid.

There was also the matter of the law. An assault charge, even if the conflict had not been initiated, called aggravated assault, could result in lifetime consequences. In law, the person being threatened is expected to explore every possible non-combat way of escaping the threat.

In our particular case then, because we were bound in by the pool area fence, we would have had to run for the fence and jump it to get away. The reality was that we would subsequently be the humiliated slaves of the muscle men.

I didn't know how Fadi and Bartholomew would react, but I definitely knew that neither Tiny nor I could accept that. If I was attacked by the very tall, big guy who stood in front of my chair, I would launch to use grappling technique to first bring him to the ground, eliminating his reach advantage, and put him in an immobilizing lock hold until the situation was resolved in a non-legal liability way, if his pride could handle it.

If not, I would hold until he tired of it all. Conversely, I suppose the big guy could eat me for dinner if I let him go. Back then, I didn't much care.

The muscle men's opening gambit was for the most muscled guy to taunt the one of us who seemed, by appearance, the toughest. Fadi, Bartholomew and Tiny were each about six foot two, and certainly Bartholomew was the largest. However, they decided that Tiny would be the toughest. Tiny did not have the oversized definition of the muscle guys, but it was obvious he was powerfully built in a football linebacker sort of way.

I guess that biggest muscle guy must have had confidence he could handle Tiny when he taunted in Tiny's direction the lame cliché, "You're lying in my chair."

To my surprise, Tiny remained immobile behind his sunglasses. I guessed he was waiting for things to develop to their full ripeness before abandoning himself to his instincts.

The guy in front of the tall but thin Fadi bleated in a sort of falsetto, "How's things, sweetheart?"

His observing no Fadi response, he extended his leg to nudge Fadi's leg off the lounge chair. It surprised me greatly to see Bartholomew suddenly raise his three hundred twenty or thirty pounds to his feet. For the moment, he removed his sunglasses and stood there defiantly as if challenging both the guy in front of him and the guy who had nudged Fadi's leg.

I was greatly impressed. Bartholomew was standing up for his buddy. I then possessed a new respect for the big guy. What other surprises would this enigmatic young man have for us?

A noticeable hesitation registered in the eyes of the muscle men, also surprise and confusion. I'm sure they were wondering if this guy had impressive muscles beneath that fat. Was he maybe a U of M offensive lineman? Was he a power lifter?

Most power lifters did not have the pumped up muscle definition of a body builder, but they were a lot stronger. These guys knew a power lifter's hug could crush them. Bartholomew was not a power lifter, but they had no way of telling.

Sensing that ripeness had arrived, Tiny launched to his feet and whipped off his sunglasses. I'm sure the muscle guys knew that a guy hyped on his naturally inspired, immense dose of adrenaline could rip all four of them apart all on his own. They were psyched.

The guy in front of Tiny extended his hands in a defensive surrender. Also in a falsetto caused by a temporary shrivelling of his testosterone makers, he pleaded, "Relax, buddy. I was just kidding."

He went on to explain that they had just opened their gym business the next block over, that they were our new neighbors at the complex, and that we should sign up for one of their fitness programs.

In retrospect, I'm very glad some legal problem did not interfere with my aviation education, that we had preserved our dignity, and that I was on the positive side of the most macho challenge I've ever witnessed. I've always wondered how Bartholomew would have performed in combat.

Chapter 17

THE GUNSLINGER

The shop portion of our one month of sheet metal classes began with the drilling of aluminum sheet metal pieces to be connected by the manual hammering of rivets against a metal block.

In a shop we would have access to pneumatic tools for riveting and such, but we needed the manual skills for when we're trying to do a repair out in the middle of nowhere.

We progressed from round headed rivets to countersinking, to countersink rivets, and the shaving of the countersink rivet heads. It was during the next, pneumatic, phase that Fadi's possible unfitness for aircraft maintenance began to be suspected.

At the beginning of the shop session after the pneumatic lecture, Fadi pushed his pneumatic rivet gun, with the air hose attached, into the waist of his pants, and, obvious to all, he stood facing Bartholomew with his legs spread apart as if he was a nineteenth century western gunslinger challenging the big guy to a gunfight.

Bartholomew just stood there staring with blinking eyes as if not comprehending his roomy's humorous intent.

Suddenly, Fadi reached for his rivet gun, pointed it at his roomy and fired. During the lecture we had been specifically warned that the gun's trigger must not be pulled if the long, metal rivet set is not positioned on a rivet head. Otherwise, the spring, even though very sturdy, could very possibly break, thereby dangerously launching the set like a missile.

I guess Fadi's mind must have been wandering during that part of the lecture. His rivet gun's spring broke, of course. The set whizzed just millimeters past Bartholomew's left ear. The would be deadly set then smashed through the shop's wall, the adjoining shop and embedded in a wooden post. We all, including Fadi, were shocked at what had happened.

In short, Bartholomew forgave his roomy's stupidity episode and Fadi survived his second intense interview with Nanny. We all put the incident behind us for the moment, but a certain underlying wariness persisted.

In the aviation world, a solidness of character was needed for people so responsible for other people's lives. Did Fadi learn from that experience which could have ended so tragically? I'll discuss it more later.

*

Our final sheet metal project was to apply a repair patch to the compound curve of a wing upper surface section. We were each given one piece of sheet metal which we were to compound bend with rollers to the right shape, and then cut a patch out of that same piece. It would have been so much easier to have a second piece of metal for the patch. We were given a second piece of aluminum, but that was for the doubler underneath the surface, which also had to be compound rolled.

After rolling his pieces to the compound shape, Fadi became very secretive about how he cut the patch out of the main piece. He bent his body closely over his piece that was covered by a large rag to furtively do his work. That naturally raised suspicion about what further nonsense he was up to.

Our project was put away in a locked cabinet after each session. That was so no one could take it away to have a professional shop do the work as a cheat. Fadi was the only one to keep his project covered at all times.

Finally, the day we were to hand in the project for grading, Fadi removed the covering rag for all to see. All in the class were astonished. The instructor said it was the finest, perfectly formed, project he had ever seen.

One had to examine it carefully to even see on the glistening upper surface that a patch had even been cut out and inserted. As well, the hand countersinking, hand forming of the flush rivets, and rivet head shaving were perfect. As with the cutting and insertion of the patch, the upper surface rivets appeared almost imperceptible.

Fadi's project was assessed the only perfect grade, in fact, the only one ever awarded at the college. Fadi looked like he had been the one who swallowed the canary, so to speak. His facial expression gave the impression he had committed a subterfuge. But how?

The speculation began. Maybe he found a way to sneak into the shop in the middle of the night, pick the cabinet lock, and substitute a professionally done piece.

As for a complete substitution, the instructor said that he had a foolproof way, using tiny etches on the underside detectable only by microscope, to ensure no substitutions were made of the original sheet metal.

Other speculations were that he had used a specially sharpened dental tool to scrape the fine cut, or that he used small pieces of soft, pure aluminum to burnish away any imperfections, including with the rivets.

Even so, that still would take considerable skill to accomplish. Fadi proudly displayed his accomplishment prominently on the living room mantle of our shared apartment.

It was amazing how such a talented person, even if it be as if a cat burglar, would later exhibit a certain character flaw regarding aviation technicalities.

Chapter 18

October and Another Incident

In October of that first year, our group went to a theater in Miami to see the movie 'Peyton Place' that had a New England setting. When the opening scene showed leaves in full color, Tiny, Bartholomew and I, all from the North, groaned loudly. We very much, for the first time in our young lives, grieved that change in color of the autumn leaves. Later, we would also even miss the cold and snow of the North. Fadi, with his being from the Middle East, wondered what the fuss was all about.

It was more than just the autumn colors. The pervasive flatness of the Florida land and the constant green of the trees' leaves irked after a while. Although I must admit that the warmer weather at that time of year did impart a certain pleasure.

It was during that October, a few weeks into our first semester that the aviation world took another serious hit – especially Lockheed again - and now directly Eddie Rickenbacker's Eastern Airlines.

Yes, another L-188 Electra crashed, the third one in only weeks in America, this time one of Rickenbacker's Electras. I could imagine Eddie's reaction, especially with the extra hit of his losing a close friend who had been the Captain of the flight.

At only an altitude of about three hundred feet after takeoff from Boston's Logan Airport, the Electra encountered a flock of about twenty thousand starlings. The prop strikes, and the engines' intake ingestion caused engine one to flame out and a temporary loss of power occurred with engines two and four. Engine three apparently stayed at full power.

The sudden slowing of the plane and asymmetric thrust sent the Electra into a nose down spin that I'm sure even Sully of miracle on the Hudson fame could not have controlled. The plane came apart on impact. Ten of the seventy-two on board survived. As tragic as it was, the aviation world breathed a certain amount of relief that another L-188 wing had not ripped off.

Two main aviation safety measures resulted from this accident. One was bird control around airports. Starlings weigh only about three ounces compared to the ten pound Canada Geese that Sully encountered, but, unlike Geese, starlings flock together in enormous numbers up to about 1.5 million as mainly a protection from larger predators. They are famous for bringing London, England's Big Ben to a stop before control measures were taken.

As one future control measure around Boston's Logan Airport, the practice of dumping food wastes in the surrounding waters was stopped.

Again unlike with Canada Geese, starlings are native to the area. The Geese Sully encountered above the Hudson were just migrants passing through the area.

The other future safety measure had to do with the passenger seats. The detached seats in this crash floated, but did so with the strapped-in , unconscious passengers face down. A redesign of the seats was obviously needed.

The incident was another entirely appropriate opportunity for an aviation safety lecture at the college. It would further sensitize us to the fact that it's not just the victims that suffer in a crash, that most of them have families and friends who also experience an intense pathos.

I've wondered how the incident with the starlings would have resulted if Eastern's Eddie Rickenbacker had had his way and used only piston engine propliners for Eastern. Would the engines have maybe at least been affected symmetrically? I wonder if anyone knows the comparative probabilities.

Chapter 19

Holiday Indignations and One More Incident

At the end of the first semester, most of us at the college dispersed to our homes for the X-mas holidays. Fadi flew back to the Middle East and Tiny was anxious to reunite with his love, but Bartholomew curiously wanted to stay in Miami.

As if by instinct, we sensed that we were not to ask Bartholomew about his personal life. Fadi was the closest to him, but he didn't know why the big guy was staying behind either. We assumed his parents had decided to visit Miami as a holiday, and that he would otherwise be involved in his church activities.

I boarded a Constellation that was to stop in Philadelphia, then at Idewilde in NYC, and on to Chicago where I would board a bus for the final leg, a long, slow leg through three states, until finally reaching home in the Sault.

It wouldn't exactly be like riding in a wagon train to the Wild West, but it was going to be a long sit. My being only eighteen years old would help.

The week before the holiday, I received a letter from my father advising me that I could again have the use of the MGA for the rest of my stay at the college.

The young fellow who had supposedly bought the car told my father that he did not qualify for a bank loan and persuaded him to let him make payments directly to my father each month which was to pay for my college expenses.

It was a sucker deal, of course. No payment was ever made. The beat up car was found in a farmer's field half submerged in mud. The young man had prudently left town. My parents decided to restore it to at least reasonably driveable condition and let me use it again for a while. The car could be sold after college when I had a job. The money then would go to cover the college cost – better late than never. In retrospect, I should have bought that great, soon to be classic car, from my parents.

I earned my gas and entertainment money working part-time for National Airlines cleaning airliners between flights. My student visa allowed that.

Tiny went home on a different flight, but since his home base was also near the Great Lakes and in my direct car path to Florida, I agreed to pick him up for a shared ride back to the college.

*

It would have been great to enjoy a quicker way home such as a Delta DC-8 flight to Chicago followed by DC-6 or 7 connection to Michigan's Kincheloe Air Force Base near the Sault, but a charter flight and the bus ride would be cheaper. It would also be my first ride on a fantastic Fat-Boy-like, rumbling Constellation.

My enthusiasm only slightly dimmed when I noticed all the military people with their very heavy carry-on gear boarding the very packed plane. It looked for sure like the plane was overloaded.

The throbbing growl of the powerful, turbo-compound, supercharged, Curtis Wright engines anaesthetized all fears. I pictured hundreds of Fat Boy Harleys tethered to the engines like teams of horses carrying us safely into the sky.

I never slept at all during the time aloft. I was having too much fun enjoying the soothing rumble that would never be experienced aboard a jet. In the darkness of the pre-dawn takeoff, the blue exhaust flames hypnotized me.

I saw the snow line somewhere in the Carolinas. It reminded me that in the excitement I had forgotten to bring along a winter coat – dumb teen. Chicago and the Sault would be well below zero.

*

We made a stop in Philadelphia, refueled at NYC's Idewilde, and took off for Chicago. During that time I was completely unaware of a horrendous event that occurred that very same day.

As we were flying toward NYC, a United Airlines DC-8 was inbound to NYC's Idewilde (Now JFK Intl.) from Chicago's O'Hare with eighty-four people aboard. At the same time, a TWA Super Constellation was inbound to NYC's La Guardia from Dayton, Ohio with forty-four on board.

On that snowy, rainy, misty morning the two airliners tragically collided about a mile away from Staten's Miller Field. One of the jet's engines hit the Connie, ripping the engine off and breaking the Connie apart.

The Connie plummeted into Staten Island. The DC-8 stayed aloft for a bit, but then crashed into some residences in Brooklyn. It was said that the pilot lifted a wing to avoid hitting a school. Six people on the ground died and an eleven year old boy was ejected from the burning jet onto a snow bank. Local people rolled him in the snow to extinguish his flaming clothes.

The boy was conscious and talking when taken to a hospital. Unfortunately, the boy had inhaled the hot gases of the burning jet fuel, which seared his lungs. He quickly developed pneumonia and died the next day.

Back then the med people were unaware of the lung damage. If they had known, an immediate application of antibiotics might have saved him.

*

In this section, I'll explain the reasons for the crash, and try to make it understandable for non-pilots and those not aircraft electronics informed.

The TWA Connie flight went normally until the tragic moment. The reasons for the accident had to do with certain developments during the UAL DC-8 flight.

The DOT report from back then is online now. We can read the facts and see the unknowns. The best way to understand why the accident happened is to think about those facts and unknowns, and place them in the context of the era.

I mentioned before that a major theme of this book is the period of aviation when piston engine propliners were being phased out in preference for jetliners. That was a factor here.

I think it best to describe the United DC-8 flight chronologically. You might remember that I wrote earlier that I had been advised it would be a waste of time and expense in my young era to pursue aviation pilot employment because the market was flooded with well-trained pilots from World War II, unless one had some kind of special influence, which I did not have, although I was counting on a technical qualification to help with that.

The United Airlines pilots involved in this incident were examples. They both, being in their forties, were veterans in the middle of their careers.

The Captain was 46 with over nineteen thousand hours of flight time. The first officer and flight engineer had logged similar hours.

What's relevant here is that the vast majority of their airline flight experience was on piston engine propliners from the DC-3 to the DC-7.

United had been a launch customer for the DC-8 the year before. The Captain had only 344 hours DC-8 time after qualifying on the type six months previously. I mention that to bring attention to the fact that his experience, not only on the DC-8, but jet flying at all was only 1.8 percent of his total airliner experience.

Airline pilots from that era would have to be the ones to comment on how the increase in airliner speed, especially in congested approach areas like NYC, affected their psychomotor ability to act quickly at those quicker speeds, and again especially in situations when Air Traffic Control (ATC) issues a sudden change in course and one of the navigation instruments is faulty, as happened in this case.

How long does it take for the brain neurons to increase to a level needed to adequately handle the quickness of mental calculation in those situations?

I also think fatigue was a factor here. The United DC-8 flight crew left Los Angeles bound for Chicago at 3:20 AM. That meant their day began hours earlier with tasks like toiletries, breakfast, commute, flight prep, and so on.

We now know how our body's circadian cycles, our sleep-wake cycles, can negatively affect our body and mind in terms of our ability to function, especially when under pressure.

I wonder what their sleep/wake cycles had been prior to that day. Had it varied much? How much truly restorative rest had they achieved before and during the flight?

With disturbed patterns, one may experience what appears to be sleep, but it's largely beta rest, and much less the restorative delta deep sleep. Chemical aids to sleep reduce the delta rest even further.

They had a two hour stopover in Chicago. The flight attendants went off shift there – lucky them. A new set of FAs came aboard. The flight from Chicago's O'Hare to the NYC area took them somewhat south to Pennsylvania where they then proceeded North-East from a radio navigation point along an electronic highway in the sky called a VOR.

*

I mentioned above that I would try to make reasonably understandable certain technicalities to non-pilots and the aviation electronics uninformed. Hence, I'll state that there were back then mainly two types of radio navigation systems on an airliner. Satellites did not exist, so there was not the convenience of GPS so common today.

VOR is the acronym for Very High Frequency Omnirange. It's the high frequency that makes those, so to speak, highways in the sky so precise. Two VOR radios aboard an aircraft were needed to establish exactly where the aircraft was at any time. VORs are still used by many aircraft today, but I'll use the past tense because it was the main system used in that era.

One of those onboard VORs was used to keep the airliner on the highway, and the other, not on the highway, was used as a cross reference so the pilots could plot on a map exactly where they were, or how far they had flown on the highway.

A problem naturally occurs if one of those precision VORs fails. The pilots will know they are on the highway, but they wouldn't know exactly where they were on that highway. No doubt you've guessed that one of the VORs failed on the UAL DC-8.

The pilots then had two choices. They could perform the tedious task of dialing the radio tuner to the cross reference VOR station for a quick location reference, and then dial back to the highway station to make sure they were still on that highway, and then do it over and over again.

The other choice was to make use of ADF. I'll explain that next and then get back to the flight.

*

The acronym ADF stands for 'Automatic Direction Finder'. Each of the two navigation radios can also be tuned to a much less precise ADF station. Please note that I wrote 'much less precise'.

ADF was an older system used before VOR was invented. It operates at a much lower frequency than VOR, takes much pilot skill to determine a flight path to a destination, and is much more susceptible to radio signal interference.

Pilots only use it if nothing else is available. Something is better than nothing. An ordinary, commercial radio station could function as an ADF. Pilots could enjoy listening to broadcast music as an undetected cross wind blew them into a mountain. A VOR would have detected the wind error.

*

Getting back to that ominous December day in 1960, the DC-8 proceeded from the Allentown, Pennsylvania VOR toward the Robbinsville, New Jersey VOR where they were to fly the Victor 123 electronic highway to the North-East.

I'm not sure when the second onboard VOR radio failed, but it was as they were approaching Robbinsville that one of the pilots contacted the United Airlines maintenance facility to inform them of the problem. Maybe they were hoping that they would receive a brilliant fix reply, but that didn't happen, and they did not inform Air Traffic Control (ATC) of the problem. I guess they thought they could handle the situation on their own without radar assistance.

Four minutes after the UAL crew radioed the United facility, NY Center surprised them with a sudden change of direction. That change would have them bypass Robbinsville so that they would intercept the electronic highway V123 earlier, shaving eleven miles off their route, which gave the crew less time to adjust.

NY Center was just trying to expedite traffic, but maybe would not have issued the change if they knew of the VOR failure. Given the direction change, the UAL would have had no problem intercepting V123 and staying on the path.

However, a big problem occurred when NY Center advised them to enter a holding pattern at the Preston intersection.

*

At this point I should briefly and simply explain some relevant things about Air Traffic Control (ATC). When an airliner is enroute to an airport, it is in communication with what's called a Center for the area the Center controls. On long trips, the airliner may pass through various Centers. The one controlling both the DC-8 and the Connie as they approached the New York area was called, appropriately, New York Center.

When an airliner gets close to its destination airport, a Center will pass control to the respective airport Approach

Control and discontinue any further contact and radar with the airliner.

In this case, the Connie came under the control of LaGuardia Approach, and the DC-8 came under the control of Idewilde Approach (now JFK). Each controller at the two Approaches could see both airliners on their radar, but only be in voice contact with the airliner destined for a landing at their own airport.

Another important fact is that those controllers could only see the altitude of the plane destined for their airport, but not the altitude of the other plane. That should have been no problem as usual because they were to have three miles horizontal separation between all planes.

That was why New York Center, before discontinuing contact, directed the DC-8 to enter a holding pattern at the Preston Intersection. At that electronic point in the sky, the DC-8 was supposed to turn 180 degrees to fly back in the direction from which it had come for one minute, then turn 180 degrees to fly toward their destination for a minute, and then keep up the same circling pattern until Idewilde Approach could see the three mile separation of all aircraft and that it would be safe for the airliner to continue toward the airport.

Both the Idewilde and LaGuardia Approach Controls must certainly have become very nervous to see that the DC-8 had flown past Preston without entering the holding pattern.

*

Once on V123, the DC-8 crew had to figure out exactly where they were on that path. The use of only the ADF as the secondary nav reference involved too much mental calculation in too short a time for those fatigued minds. They had also been directed to slow to around 250 knots.

This would be the very first time a flight data recorder would be used in a crash investigation. So the investigators knew they had been flying at 301 knots at impact, which left the crew even less time to adjust. The Connie did not have a flight data recorder, but since it was in a decent, its airspeed would have been around 190 knots.

When LaGuardia Approach saw that the DC-8 had blown past Preston, they did not know that jet's altitude, but were certainly alarmed, so gave a direction change, told them to keep descending and to watch for traffic within one mile. Visibility was only one mile. We don't know what either crew observed outside their windshields.

When the DC-8 crew advised Approach that they were approaching the Preston Intersection, they had already flown past it. They were about 12 miles off course when the collision happened, as described above.

At that awful moment, I was in another Connie somewhere in the Carolinas approaching the very same airspace. With we passengers being mercifully unaware of what had happened earlier, we landed at LaGuardia, refueled, and took off for Chicago.

Chapter 20

To Home and Back

After my disembarking from the Connie at Chicago's O'Hare, a man in the terminal who had been scanning the passengers settled his gaze on me. He asked me if I needed a taxi ride to the bus terminal. I did indeed, and so did three, young like me, military guys.

When we arrived at the bus terminal, the taxi guy unexpectedly charged each of us the full fare on the meter. So his scam was to get four naïve fares for one trip. That would very well make up for his loss of the usual tip.

Maybe he had twelve kids to support, or maybe he was actually a multi-millionaire unable to control his compulsion for the hustle.

One of the military declared through clenched teeth, "You're only getting one fourth the fare from me."

The taxi driver had chosen his parking spot with tactical precision. He pointed to a policeman seated on a horse close to the taxi and stated with a sardonic grin that begged to be crushed, "I'll tell the policeman right there that you refused to pay the fare."

The military guy laughed. "I'm not refusing. I'm offering to pay my fair share, you greedy guts."

The taxi guy, almost drooling at the high he was getting off the incident, responded with his grin now absurdly contorted, "I'm sure you'll win your case in court. Meanwhile, I hope the bail and lawyer fees won't be oppressive. I would hate to see your wonderful holiday with the family get messed up. And I'm sure the military is very understanding about those who report late."

To the accompaniment of a full set of loud, awful, even sinful words bouncing off the taxi's interior surfaces, we each fished for the cash. The taxi man then actually drooled as he received the currency.

The taxi driver's evil grin sometimes pops up disturbingly in my dreams. Is that PTSD?

It was a long, tiring bus trip up Lake Michigan through Illinois, Wisconsin and Michigan to the Sault. The trip from Miami had taken at least a full twenty-four hours.

It was great to see loved ones, to put on a warm coat, to crash into bed, to have the use of the MG again, and to be a bit more worldly wise.

*

The MG looked quite beaten up, but I was glad to have it. What I did not see was that the car's mistreatment had caused in it a negative psychological shift. It had become evil, but like any patient predator, it would craftily wait until I was at my most vulnerable before attacking.

Also, it was greedy like the Chicago taxi driver had been. It wanted at least two victims in each attack. It waited patiently for Tiny to come aboard.

Tiny's home was conveniently in a town directly on my path south. The MG wasn't happy that I began the trip so early in the morning. It preferred darkness to unleash its attacks on the two of us.

The sun was up by the time I reached Tiny's place. Considering the cheap ride to Miami, I thought he would greet me in a buoyant mood. His buoyancy would be inspired by his getting on with his much desired aviation education, a joyous family reunion, a joyous reunion with his beloved, and the prospect of a shared expense ride to warm Miami in a sports car. However, Tiny, his parents and siblings, all seemed a touch reserved.

My mind speculated. Had there perhaps been a death of a close relative or loved pet? Had his relationship with his parents deteriorated? Then it occurred to me that Tiny's deeply beloved girlfriend wasn't there to see him off. It seemed absurd a guy as handsome as the finest movie stars would be dumped.

That was in the days before I became familiar with the biographies of some of those movie stars which revealed their wretched dumpings and futile hunts for true love.

As the trip progressed, Tiny gradually revealed the details of what had happened. I knew the MG was listening in on the conversation. I heard a distinct purr from the engine's sick, pathological enjoyment of Tiny's agony at the point where Tiny wretchedly revealed how he had completely given his heart away to his beloved to whom he had been so faithful.

As Tiny's mood shifted from profound sadness to profound moroseness, the MG lost its patience. It preferred to strike with its nastiness in complete darkness, but the time had ripened.

A bit before sundown of that day, we began to smell a whiff of smoke – burning wood smoke. At first I wondered if there had been a forest fire or some other kind of burning in the area. We did not see any smoke outside.

Looking back on it, I remember the engine noise had gradually, almost imperceptibly, grown louder. The exhaust pipe had opened a small hole on the upper surface that blew very hot exhaust directly at one panel of the MG's wooden floor.

The panel suddenly caught fire, revealing a flame and a lot of smoke under Tiny's feet. I quickly pulled to the roadside where we jumped out. Tiny reached in to rip the carpet away and extract the burning panel. He stomped out the flame. So what to do then?

I wasn't going to leave the car behind, but the prospect of driving it as it was presented certain difficulties. Just the smell of the charred wood might be okay, but the fumes of the burned carpet had deposited all over the car's interior. The smell was awful and I'm sure carcinogenic, although I had not known the word back then.

The exhaust leak would continue to shoot upward, and a cold blast of freezing air would also blast up through the gaping floor hole directly at Tiny.

Tiny threw up his arms, emitted some kind of evil pronouncement against the world, and declared that he wanted to hitchhike to a large airport where he could catch a flight to Miami.

My solution was to get my polar sleeping bag out of the trunk so Tiny could get in it and ride warmly, snugly in the passenger seat for what I was sure would be a short drive to a town just a couple of miles up the road that had a local airport large enough to have an FBO (a Fixed Base Operation that did various types of general aviation business).

I was sure they would have welding equipment we could, hopefully, freely use to plug up the exhaust leak. Until then, I would also get the heavy blanket, also stored in the car's trunk, to wrap myself in so we could leave the MG's convertible top down and the floor panel stored behind the seats for the short jaunt.

As opposed to how skimpily I had been dressed for the trip north, I now had a good, winter, hooded parka, and so did Tiny. We would have to leave the top down, of course, because of the foul smell and avoidance of exhaust asphyxiation. I should explain that from where I came from in Canada, it was standard practice during winters to store a polar sleeping bag, a heavy blanket, a bag of dirt for traction in snow, a small shovel in one's car, and a survival kit with candles, matches, beef jerky, chocolate, liquid and so on.

Tiny reluctantly agreed to give it a try.

*

Naturally, it began to snow harder. Tiny climbed into the sleeping bag and sat with his long legs extended over the transmission hump. He looked spooky with the sleeping bag's top over his head and with his peering out through only a slit.

I was able to grasp the steering wheel, operate the pedals and the four speed manual transmission adequately with the blanket wrapped around me and the parka.

With the top down, the floor panel stored behind the seats, and the snowflakes swirling around us, we headed toward that nearby town.

I'm sure the town's citizens on the main street must have observed us with a mixture of oddness and possibly even fear. The Canadian license plates would also add a touch of the exotic.

I stopped at a store to buy a polar sleeping bag for myself. Because of the MG's interior smell, we would probably have to drive the rest of the way to Miami with the top down. At least the air temperature would gradually warm.

*

The sun was down when we reached the airport. Although the airport runway lights were on, I was disappointed to see that the FBO was in darkness. I guessed everyone, as would usually be the case, had gone home for the evening.

A trailer with a dimly lighted window was located beside the hangar door. I parked in front of the trailer door, got out, and knocked.

A very broad guy with a stringy beard opened the door, and just stood there staring at me without expression. I guessed he was the FBO's afterhours watchman. Had he just been released from incarceration for having done something really awful and disgusting? Had he perpetrated other heinous things that had not yet been discovered? Was that the smell of boiling homo sapien flesh wafting out at me?

I explained, with the odd pathologically induced stutter, as best I could, our predicament. When the chainsaw serial murderer continued to just silently stare at me, Tiny felt the need to make his presence known.

Rising up with his feet planted on the passenger seat, sleeping bagged Tiny loomed large like an enormous cobra. The grizzled chainsaw man reached inside his coat. Did he have a gun? Maybe Tiny's presence in the scene wasn't a good idea. Too late.

Tiny unzipped his sleeping bag, stepped out of it, and lowered his feet to the ground. He tried to appear as unintimidating as possible.

Chainsaw pulled out a big set of keys and barked, "You can use the oxy-acetylene torch," as he proceeded to the hangar door. "Cost you ten bucks."

We restrained ourselves from lecturing him about his ungentlemanly manners and greediness.

Tiny welded the exhaust pipe damage, after which, we placed the charred panel back in the floor opening. The panel had a gaping hole in the middle, but the blast of cold air would be greatly reduced.

We climbed back into our sleeping bags, which were surprisingly warm and snug – like a womb. I had no trouble grasping the steering wheel, shifting the gears and operating the pedals from inside the bag.

We set off down the highway, hoping with a diminishing confidence that the MG would behave the rest of the way.

*

As we headed further south, I thought it was generous of the MG to give us several hours for Tiny to get a good sleep before any more nonsense. Hence, I've downgraded my almost anthropomorphic assessment of the vehicle from evil to simply negligent.

Tiny seemed to be not disturbed at all by the cold air and snowflakes swirling around his sleeping bag cocooned head.

The swirling snowflakes would ordinarily have induced drowsiness in me too, especially as the early morning hours droned on, but I had taken a special precaution.

During one fuel stop, I noticed displayed on the counter a product called 'No Doz'. The clerk said that transport truck drivers and airline pilots used those pills all the time to keep awake and alert during their 'red-eye' schedules.

I bought a package, popped one pill, and then, without reading the recommended dosage, my teenaged brain reassured me that a second pill would be even better. I immediately swallowed a second one.

I was totally unaware then that the pills were amphetamines that would later become illegal, and even race hearts to a point where sudden death sometimes occurred.

I must say that 'No Doz' got me through the night very well, without any tiredness at all, but I did not like the intensity effect on my body and mind. I'm sure the intensity would have been less had I not taken the extra pill, but I don't know exactly how much because I never again felt compelled to take an amphetamine.

*

At about four in the morning darkness, the MG again became playful. It shut down the engine. Tiny stirred groggily as I pulled over to the side of the road.

Disoriented, he asked what was going on. I told him about the failed engine and added that it had happened before and that I thought I knew how to fix it.

It was when I tried to turn the ignition back on that my suspicion was confirmed. I did not hear the familiar click, click, click from the electric fuel pump.

I should have known that the pump's electrical points must have deteriorated during the long abuse, making them susceptible to sticking. I should have filed the points smooth before embarking on the long journey south, but the holiday excitements caused me to forget.

I'll explain that electrical points are used in certain electrical systems to create the very high voltage needed for such things as igniting a spark plug in a piston engine cylinder and driving, in this case, a fuel pump that is not mechanically driven by an engine, such as in the MG.

Please note that some aircraft use such electrical fuel pumps with points, but only as an emergency backup. Such pumps must be inspected at regular inspection intervals.

Those inspection intervals are mainly selected to prevent one of the pump electrical components called a capacitor from failing. The same as in an engine ignition system, the capacitor is wired in parallel with the points. When the points open to build the high voltage, the current wants to keep going, but is usually absorbed by the capacitor. If the capacitor fails, the current causes arcing and pitting of the points and they stick. I obviously needed to replace the pump capacitor.

I had in my tool box in the trunk a fine file to dress the points, and a gauge to set the gap, but had no spare capacitor. The situation called for a very quick fix.

I opened the MG door and stepped out to go back to the car's trunk to get at my tool box, but the 'No-Doz' had caused my legs to become very weak and very wobbly. I staggered uncontrollably to the center of the highway and collapsed, laying there unable to get up or even crawl.

I had been able to operate the car's pedals okay to that point, but mobility beyond that was really something impossibly else. Did transport truck drivers and airline pilots encounter the same problem? I'm sure they all took the recommended dosage.

"Tiny?" I called.

"What?"

"I'm collapsed and paralyzed in the middle of the highway. You can either help me back into my seat or take the car and leave me as road kill."

Tiny quickly helped me back to my feet and guided me as though I was a stumbling drunk to the passenger seat.

"You're riding shotgun from here on in, roomy," he advised me.

I reached over to turn the ignition on and told Tiny to go back to the car's trunk and extract the rubber mallet from my tool box, then lift a panel out of the way from behind the seats and pound the fuel pump until it started clicking.

Tiny pounded and the pump click, click clicked. He then climbed into the driver's seat, roared the engine to life, and we were off again. I would replace the points and capacitor later in Miami when I had more strength and sanity. Meanwhile, we kept the mallet handy.

*

Just before dawn, the MG decided to be naughty again. In the previous couple of hours it had gradually become resentful of the fuel pump maintenance neglect and distrustful of my good intentions.

In the complete darkness, the headlights failed first, and then the engine quit. I knew it was not the fuel pump misbehaving again.

The 'No-Doz' had prevented my falling asleep, but my apparent alertness had been a subtle deception. My chemically induced, rapid succession of delusional fantasies had prevented vigilance.

My scan, at regular intervals, of the MG's ammeter and engine temperature gauges had become impaired. I normally should have noted that the temp gauge indicated a normal reading but that the ammeter (electrical current) had progressively deteriorated to zero.

Even though the ambient air temperature flowing over the engine was cold, the many hours of driving had overly heat soaked the engine and its attachments, thereby causing the electrical regulation systems to deteriorate. The generator no longer charged the battery, causing the headlights to fail first, and then the ignition system. The engine's spark plugs no longer sparked.

In complete darkness without the lights, we departed the highway and swirled around in the snow beside the highway. I sank low in the seat and held on. Fortunately, thanks to the MG's low center of gravity, it did not flip. We survived, but we were stuck in a snow bank. My little bag of dirt and small shovel in the trunk were inadequate for that situation.

Tiny left me sitting there and hitchhiked to a service station a couple of miles away, returning after a while in a tow truck. By the time the battery had fully charged at the service station, the sun was nudging the horizon. That meant we no longer needed the headlights, but that we needed to find a British car dealership for either a new electrical regulator or alternator or both.

Fortunately we were only about an hour outside a major metropolitan center. The service station guy generously looked up the address of the British car dealership and drew us a map of how to get there.

The reduced electrical demands of our four cylinder engine ignition allowed the charged battery to get us there unmolested. Life was sweet for at least a bit, but the MG wasn't done with us quite yet.

*

Not long after passing over the snow line, the weather became somewhat warmer. The same as with Fadi's Caddy, Tiny enjoyed driving the MG. After a while, we were able to climb out of our sleeping bags, although we still needed the parkas for a bit yet.

Travelling into an area that had just shortly been inundated with heavy rain, my fears about the MG's fragile Lucas electrical ignition system were induced. It didn't take long to happen.

After negotiating a corner, we unavoidably hit a large puddle. A spray of very cold water shot up through the burned out hole in the floor panel and soaked me right up my middle. It was lucky for Tiny that I had taken the 'No-Doz' or he would have been the guy soaked. The MG's engine naturally quit instantly. We pulled to the roadside.

My first fear was that the ignition's distributor cap may have cracked because of the cold water's shock cooling of the heat soaked distributor.

With my legs 'No-Doz' toasted, I suggested to Tiny what to do, because it had happened to me before.

Tiny got the tool chest out, along with some rags I kept for such occasions and opened the hood. He took off the distributor cap and disconnected the ignition leads from the spark plugs.

Luckily, the cap had not cracked, but the shock cooling had caused the distributor and its leads to plunge below the dew point. Everything was condensation soaked inside.

Since Tiny could use his legs, he wipe dried the distributor body and plugs, and I, in the passenger seat, dried the distributor cap and leads.

It was a time consuming inconvenience, but better and cheaper than alternatives. If I had felt bodily capable, I would have adjusted the two carburetors as well because the fuel/air ratio was running lean due to the warmer, ambient air. The lean mixture would cause even more heat soaking on this long drive.

Setting those carbs was an art form. One carb fed more mixture to the cooler front cylinders, and the other one fed more to the hotter rear cylinders. The mixture setting was therefore different for each carb.

I adjusted them according to the engine's sound and RPM. I performed the function of later electronic sensors. Cars were a lot more fun back then, unless you're stuck on a highway with 'No-Doz' legs and a soaked crotch.

The car started beautifully. Such success made life seem giddy wonderful, at least for the moment. Soaked, I managed to get back into my sleeping bag. We drove on.

It was just outside Tallahassee in northern Florida that the final problem hit. The steering wheel seized, but Tiny's strength made it move just a little. So we puttered along the roadside at almost a walking pace until we reached a service station.

I thought my dad had at least taken the MG in for a lube job after finding it abused in that farm mud. I guess not.

We got the lube, the steering loosened, and we continued on, buoyed by the warmer weather and the prospect of soon finishing the odyssey.

Finally at Miami, Tiny helped me into the apartment and shoved me onto my bed. In the coming days I would have to perform a complete cleaning of the MG's interior to get rid of the smell from the floor fire's fumes. I would cut a new wooden floor panel, scrub the seats, all exposed surfaces and each individual wire and cable behind the instrument panel.

Chapter 21

Rosemary and FAs

I was having what could later be characterized as very close to a lucid dream, a dream in which we're aware we're dreaming. In a real lucid dream, we can interact with the characters and environment given to us in that dream world and create our own scenarios.

In my case, I seemed to know I was on my bed in Miami at the Silver Springs complex and couldn't move. I could only listen to Tiny's words directed to me. He was sitting on his bunk, facing me.

"Bartholomew got married during the holidays," he said. "He's now living with his new bride a couple of apartments down from us. He says he'll keep paying his share of our rent until we can find another roomy replacement for him.

This is absurd, I told myself. I know I'm dreaming. I'm dreaming this because of the 'No-Doz' as a side effect. It was playing with my mind.

"Bartholomew did go home for the holidays a couple of days after we all left", Tiny went on," but had trouble with his parents when he told them what he was going to do. So he came back here after only a couple of days and they married in Miami quickly with a really simple, secular ceremony conducted by a Justice of the Peace.

I was pulling out of the dream now. Though still hazy, I seemed to know that Tiny really was sitting there and saying those words. It seemed like his eyes were very wide with surprise as he spoke.

"Fadi is upset that he wasn't best man, but he had gone back home for the holidays like us. Her name is Rosemary.

She's from that ecumenical church he was attending in downtown Miami on Flagler. Fadi says she's about ten years older than the big guy."

What immediately came to my mind was the Nun Bartholomew had waltzed with at the Barry College dance. But that couldn't be because the woman he married had been attending his church.

A Nun wouldn't do that, and it's an absolute law of nature that a guy join the church of his wife. It's never the other way around, or so my young mind thought.

"Fadi says she used to be a Catholic," Tiny continued. "He says that Bartholomew quit going to his fundamentalist church and started attending what they call an ecumenical church – one that anyone of any faith can attend, even non-Christians. I guess she popped his wheelie, and he popped hers."

Although this was truly amazing, even shocking news, I concluded that Bartholomew's bride, her being older and a former Catholic was mere coincidence, and that she was not that Nun from the Barry Dance.

Fully awake at that point, I asked Tiny anyway if he knew if Rosemary was the Nun. He merely shrugged, and added that Fadi had also only shrugged when he asked him the same question.

When we're only eighteen, such things are adopted seamlessly into our universe. The mystery remained. We replaced Bartholomew in our apartment with a business pilot student who had been staying at the usual Flagler Street hotel.

When I did see Rosemary, she did, in fact, look very much like the Barry College Nun, auburn hair and all. She looked quite attractive with her hair hanging loose, and with her conservative makeup applied, just like I had imagined at the dance.

We all noticed that Bartholomew was quickly losing weight, and he now made eye contact with people. It was as though an enormous, psychological burden had been lifted off his shoulders.

I was happy for him, and for her too.

*

Tiny was anxious to help me clean the smoke deposits in the MG because he wanted to borrow the car for a date he arranged with a Barry College student the first weekend after our return from the holidays.

At the time, I was surprised at his very quick recovery from losing the devotions of his, until the holiday, eternal love of his life. If I had been at a late stage of my own life, I would have reasoned on the validity of a cliché like 'getting back on the horse if bucked off' as a helpful antidote to the immense psychological pain of losing such a beloved one.

But I would have also considered the often enormous effect it would have on the new love interest. What about the rebound effect?

People with crushed hearts are usually counselled to somehow endure a cooling off period in fair consideration of the new person's feelings. It's been said that rebound love is pseudo love for the rebounder, but very real for the reboundee.

Some who have suffered rejection say they would rather have been subjected to the tortures of the Inquisition (if not completely thought through).

I told Tiny that he could use the MG that once, but that I would help him find a really cheap, but repairable, Junker at an auto junk yard that we could salvage with junk parts.

He could thereafter park it, hidden in the trees, opposite the Barry College main entrance when he went over to pick up his new dating person. Tiny would date that person he had met at the first Barry dance every weekend for the rest of our time at Embry Riddle.

I late peeked into Tiny's dresser bottom drawer. The photo of his old girlfriend remained unmolested.

Fadi also regularly dated the Barry College young woman with whom he had ballroom danced. He had no problem parking his gleaming Caddy right in front of the dorm.

I'll comment further on those relationships later.

*

There was no exact point when the gender specific terms stewardess and steward were replaced by the gender neutral term flight attendant.

In fact, there are places and airlines in the world that continue to use the old terms. Some use other terms.

Stewardess and steward were the terms mainly in use when I attended Embry Riddle, but I'll simply use AF to refer to them here. If I need to differentiate the genders, I'll indicate gender difference by using the terms female or male before FA.

Both Eastern and National Airlines, with both their headquarters being in Miami, housed their trainees just up the street from us in a couple of other complexes with a similar set of apartments with a large open area around large swimming pools.

The existence of those primarily female FA domiciles near the Embry Riddle males caused a dynamic tension to exert its awesome strength, but it differed from the situation with Barry College.

I'm sure the FAs did not conspire to deliberately set what I'll refer to as the FA domicile trap. The trap developed quite all on its own due to the natural flow between the opposite poles of humanity and our various pack mentalities back then.

As we all know, all of us are truly sweet, harmless creatures if not warped beyond belief. It's only when we relatively unwarped creatures pack together that we, as if by magic, transform to people who become willing to approach the dark side.

Some would say that it was almost an inviolable law of nature like when lions and other pack predators victimize the weak of a herd. They say that the culling even strengthens the herd.

Curiously, the weak in the Embry Riddle herd were the narcissists among us. Most know that a narcissist has an excessive self-regard or self-love. The condition becomes quickly evident in one who can only talk about one's own fabulous, personal qualities and interests.

Absolute proof is when a narcissist becomes aware a mirror is near. The narcissist will then disappear for the rest of the day.

We had a set of narcissists in our class. Everyone knew they had to be culled. It's interesting that such ones will form an exclusive group since they will only talk about and consider themselves. I postulate that they see themselves reflected in each other's faces. They transform into a collective, as if they were one narcissist.

Their words of self-love when grouped are like pixie dust. One sprinkle and poof – magic happens.

Women can be narcissists too, of course, although it's expressed differently, but we will never see a female FA narcissist. FAs truly have to be interested in other people, even to the extent of bravely disregarding their own safety in an emergency situation.

Aviation techs (pilots too) must also be interested in others. They have to do those extra things we were taught to do so those big airliners with their friends and families aboard will arrive safely. It's why the cull was necessary.

I think it was a member of the Embry Riddle staff who informed the narcissists about that available apartment at the Eastern FA complex. I saw him one day in shop penetrate a pixie dust session. He actually got them to listen to him.

That pixie group left the Silver Springs for the apartment in the FA complex that very same day.

Very quickly we learned that the FAs turned the guys into their slaves. The FAs shared their new slaves among the apartments to have them do their laundry (nicely folded), wash their dishes, and any other non-intimate things.

It was fun to hear the pixies complaining about how none of them could get even one date. I think the reason they endured the situation so long was that they thought it impossible for their person and charms to be disregarded. They were blind to the evidence right in front of them, from day to day to day.

You might reason that the pixies would have been disappointed and disillusioned with their nearly shattered psychological mirror, but it wasn't exactly a cull. Since pixies can't be cured, wouldn't we just end up with sad narcissists who work on airplanes?

Not so. The reason is that the FA domicile trap, as if magically, causes the narcissists to leave the program, although, the regular safety lectures, which also acted as sensitivity training, would take them painfully into the realms of people other than themselves. That one-two punch must be too much for them. It was a righteous cull. Aviation needs solid people.

I believe that the Barry College people would have set up a Barry cull very similar to the FA one if they could have. All humanity are instinct proportional, especially in packs. The only difference is that guys could not move into the Barry dorms.

Chapter 22

Rollo and his Cloud

I won't reveal who this instructor was, or what subject he taught, but he was definitely not capable of the fine quality of instruction usual for the college.

To instruct properly, one must engage the student's attention, and reveal to the neophytes the great techy secrets one expected from a college with such a great reputation. The boot salesman instructor had been eccentric, but I couldn't fault him on the quality of his instruction during his saner moments.

This instructor's one accomplishment was to have published a well-regarded aviation book. I'm sure he possessed a reasonable intelligence, but he never made eye contact. He was always at his desk in the classroom, writing on a pad of paper as he students entered the room for his class. I guess he was updating his book or writing a new one.

At the right moment, and without looking up ever, he read very slowly from the text for the whole session. I could silently read it much faster myself. There were no comments, or experiences, or extra insights into what he was reading.

When he finished at the end of every class, he would dismiss us, again without engaging our person, and resume his writing. Something had to be done about him, in a student justice sort of way.

One day one of our classmates, whom I'll call Rollo, performed an experiment during a lecture, an experiment for which he was eminently suited.

Most other people could probably eat a breakfast of pork and beans, fried cabbage with broccoli, beer and cheesecake, and suffer only a mild intestinal disturbance that could be handled with mildly silent emissions gradually spaced over time, with, perhaps even a major seismic disturbance between classes.

Rollo's intestinal flora, cultivated I'm sure by a lifetime of extreme culinary insult, was different from that of most other people.

Right after the instructor began reading from his text, Rollo let rip with a blast that would rival a hundred tubas played wetly. The green cloud spread quickly while we stumbled over each other in an attempt to clear the classroom. The gasping and exclamations of pain were pathetic.

The instructor, now the only one in the room, stayed seated without looking up. His reading continued uninterrupted for the rest of the class time in that empty, fetid classroom. I would have to qualify that as one of the most amazing incidents I've ever witnessed. As with the story about the boot salesman instructor's rabid behaviour, it's the sort of story you might think the author invented, but I assure you it happened just like that.

That instructor soon left the college. So, as with the FAs domicile trap, Rollo's cloud had the beneficial effect of one more righteous cull.

Chapter 23

The Apocryphal Tri-Pacer Story

I'm not sure if this is an apocryphal (false) story. I've not personally witnessed such an event. Maybe there was a similar, original event, and it's been told and retold with various, accumulating versions.

In this version, an Embry Riddle instructor (I'm not saying he was the boot salesman instructor) asked one of his more competent students to help him with the inspection of a Piper Tri-Pacer located at a small Florida airport very near the ocean.

Being very near the ocean meant that the humid air had a very high salt content. Knowledgeable aviation people know that salty air causes accelerated corrosion of metal parts.

Because the selling price of the Piper was quite low due to the poor plane's long period of abandonment and lack of use, the instructor wanted to buy it, restore it, and resell for a nice profit, much the same as real estate people will renovate and flip a house for profit.

The Tri-Pacer was/is a tube and fabric, small plane with tricycle gear that was more powerful, quicker, and more maneuverable than beginner planes like a Piper Cub. A pilot needed more skill to fly it.

Naturally, and appropriately so, the instructor wanted to test the plane' chromoly tube frame for integrity. The testing would involve climbing into the plane right back to the dark tail and squeezing the tubing in several places with a set of scissor type, ice picks. The pick ends were blunted with little leather mittens, of course.

A lot of other tests were to be performed that day, including other types of corrosion testing. Because of that salty air, the age of the plane, and the long period of sitting idle, the Piper's testing had to be thorough, which meant even climbing into that tight place under the instrument panel to examine all the electrical leads. Everything that moved had to be moved, with examination at every position, and so on.

The engine didn't matter at this time. The instructor was going to have a close relative at a certified engine overhaul shop do the restoration cheaply with his own participation.

Of course, the instructor wanted the student to use his young physique for the difficult inspections that involved climbing into difficult spaces, like under that instrument panel and the dark, confined, rear of the fuselage.

The chromoly steel tubing's interior had originally been soaked with oil to protect against corrosion, but, in mucho time, that oil could harden, making the tubes vulnerable, especially in salty climates.

The student then revealed very disappointing news to the instructor. He informed the pudgy, older instructor that he was claustrophobic. The instructor noticed that the student was now breathing more heavily than normal, but was not yet hyperventilating. The student, with his face becoming clammy and reddening, offered that he could probably manage under the instrument panel, or so he said, but that he definitely could not climb into that dark, tight space at the back of the fuselage.

The instructor stared at him disconsolately. This young man was apparently unaware that a big requirement of an aircraft tech was the ability and eager desire to crawl into dark, tight spaces.

The instructor revealed to the student the fact that he too was claustrophobic, but that he always was able to master that fear because he loved working on airplanes so much.

The student nodded, but remained adamant. He could not possibly do it without a total, screaming freak-out that could involve hospitalization.

The instructor, after emitting words that delicate ones should not hear, summoned all his mental strength, grabbed the ice picks vigorously, focused his quivering mind on slow, deep breathing to prevent hyperventilation, and began the crawl behind the seats into that dark pit of horrors.

The student went to the Piper's front to hold the plane's nose in case the balance became upset causing this tricycle gear plane to tip backwards, clunking the tail on the ground and maybe injuring the instructor.

No doubt you've guessed what sadly happened next. The preserving oil long ago applied to the inner walls of the chromoly tubing had indeed hardened and cracked, making them susceptible to corrosion. As well, the fabric had age hardened, especially after all those decades of exposure to the very hot Florida sun.

A coating of aluminized dope had been originally applied to protect against deterioration by the sun's ultra violet rays, but time did its nasty thing anyway. It just took longer to do it.

Another factor in the present situation was that the instructor's now hyper imagination started to take control. He knew of a case in which a deadly Florida snake had attacked another inspector who had crawled into the dark back of a plane. It had been a Tri-Pacer too.

Four things happened very quickly. The snake, imaginary or not, rose up with a menacing hiss from its wide open fangs directly in front of his flashlight. Then, under the pudgy weight of the instructor, the tubes snapped and the fabric easily ripped. As the tail solidly hit the ground, the instructor was catapulted head first into the very back of the tail where his head became wedged. The control cables held that broken aft part of the fuselage vertically.

The student was thrust into a state of shock by the sight of the instructor's exposed leg's frenzied thrashing and the screaming. The student also screamed and then collapsed, writhing on the ground.

Fortunately, someone heard the screaming. He said it sounded almost beautiful, like the Beach Boys in full harmony. The two had passed out by the time help came. The traumatized snake had prudently withdrawn. Both men were admitted to the hospital psych ward where they were treated for psychotic breaks.

It's said that the student later seriously reconsidered his dedication to the technical aspects of airplanes. It's rumored that the instructor became a boot salesman.

*

While on the subject of corrosion, I would like to seriously bring your attention to a push rod maintenance concern. Especially in salty air areas in close proximity to an ocean, the ball ends of push rods should be free to rotate.

It can be extremely dangerous for corrosion to cause the ball ends to seize. An example is with pushrod driven flaps. The plane's flaps must, of course, extend and retract symmetrically.

I know of at least one occasion for sure when corrosion had caused one of two flap pushrods ball ends to snap off when the pilot tried to extend the flaps for landing.

It was as if a giant aileron on only the one side had lowered, sending the plane immediately into an uncontrollable spin into the ground.

During inspection, we need to grab the pushrods and move them up and down to make sure they can rotate freely. We do that whether or not we are in a corrosion zone. Some aircraft employ a figure eight cable arrangement between the flaps to prevent asymmetrical deployment. Some do not.

Chapter 24

The Apocryphal Cable Splice Story

When an instructor was introducing his lecture about cable splicing, one student asked, "Sir, when will we ever have the need to do this? Splicing looks like a long, tedious process. Wouldn't it be so much easier to just replace the cable with a new one?"

The instructor bestowed on him the sardonic, weary smile of one who had to summon the mental strength to repeat once again the same information he had repeated over and over again, year after year.

"Use your imagination," he went on. "Ask yourself some questions. What do you do if there are no ready-made cables available to replace a failed cable and an important flight must go within the hour?"

Another student asked, "Sir, woven splices, depending on the cable type and size, may have only sixty to seventy-five percent the original cable strength. What do the regulations say about that?"

The instructor smiled with a cocked and raised brow as if observing an exotic insect. We instinctively sensed that such a sardonic smile, a cocked head and a raised brow meant that he was happy the student had done at least the basic technical reading, and that the question had been raised, but that he was going to play a mind game with us to reveal to our young, naïve minds our limitations as thinking human beings.

He opened his arms wide as if to include us all in the process. "Someone, please, inform this uninformed mind among us what air regulations apply here. Under what circumstances may an airplane be permitted to fly with reduced cable strength?

We all sat mute and embarrassed. None of us had researched the regs. The instructor gloated with extreme pleasure in our discomfort. This sort of shtick was one of his strategies for mental survival on a job that required him to repeat once again the twelve times a year, year after year, same information.

Repetition of the same thing time after time has an immense power to destroy an instructor's mind. There must be some variation, even comic relief. Seeing the light of understanding ignite in the eyes of the students in response to a deep insight is not enough.

It could be compared to a Broadway actor who performs the same lines in play after play, year after year, for a thousand shows. There's always some variation in how the lines are presented, in the interactions with the other actors, and, especially, how the audience variously responds to the acting. Those variations are necessary for mental health.

We, as students, would benefit from his shtick. We were then motivated to research the regs on the topic and prove it the next morning, as well as on the month-ending exam. Henceforth, we would not only become competent in the regs about cable splicing, but also in the regs relating to all the topics of the entire program.

*

With the instructor having derived his sadistic, but mind enhancing pleasure from our blank-stare, dumkoff reg stupidity, he launched into his apocryphal cable splice story.

I say apocryphal because I don't think the story is exactly true. I think he either creatively invented the story or he exaggerated a true experience.

The instructor began pacing back and forth as if on a stage. His arms and body would gesticulate in punctuating his story. He was as if a Broadway actor performing his thousandth repeat of the lines, but being neurologically nourished and restored by it.

"I took off in a float plane with a friend, Charlie, one late October day at the first hint of morning light," he began, "headed for the high Alaskan arctic for a day of what was to be fantastic fishing in a small, but pristine lake.

The fish from that lake tasted like no others in all the lakes of the north at that time of year just before the freeze over.

We had rented a Cessna 180 at an Oceanside seaport, lashed a canoe to one of the floats and lifted off with a full fuel load. That fuel load would get us to the remote lake, but there wasn't much to waste if we were to get back to the seaport that evening.

It was a beautiful day for a flight, and I could already taste the succulent fish as if I was actually eating it. We intended to cook up a good meal before heading back.

The lake was too small for a normal landing, so I side slipped the plane in to reduce the landing run. However, a big problem was that the lake was so pristine that I couldn't see the surface. I could only see the bottom.

Fortunately, I had done plenty of side slip landings and instinctively pulled out of the slip just in time. Charlie knew he was having a wild ride, but was unaware of the real danger.

We unlashed the canoe and had a great time fishing to the limit. Unfortunately, the weather was quickly closing in, so we reluctantly decided to delay the succulent fish gratification until later back at the lodge. We lashed the canoe back onto the float and made ready to leave.

I performed an external pre-flight, as usual, and then climbed into the cockpit where I tested the integrity of the flight controls. I tested the aileron on each wing for proper deflection, and then pulled the yoke back to the test the elevators.

That's when I knew a big problem had suddenly developed. Feeling no resistance on the control, I instantly knew that a cable had either snapped apart or come loose from a pulley.

It meant, of course, that I then had no way of controlling the up and down motion of the airplane. The good news was that it happened then and not after takeoff.

When I told Charlie, he stared at me in horror. We were a very long way from civilization. I could see his mind calculating how many days of walking it would take to get back to base.

We had some minimal camping equipment and some emergency supplies like beef jerky, but the grizzlies would be a big problem, and maybe wolves and the occasional psychopath too.

As for me, it was one of the happiest days of my life. For years, aviation people had scoffed at me for including a short length of control cable in my tool box for splicing in the event of such an eventuality as in our situation.

That short length of cable took up very little room in the tool box, so why not? Charlie exhaled with relief when I told him. I grabbed my tool box, put on my miners forehead lamp, and crawled into the dark aft of the fuselage to start the repair.

I imagined the giddy joy I would have each and every time I would reveal this incident to the scoffers. I almost burst out laughing at my imagined images of those scoffers eating their scoff. Life was sweet.

*

"As I worked my way toward the cable break at the back, I examined the cable mainly at the pulleys and grommets where the wear takes place.

As suspected, I saw there was wear beyond limits. The cables were of stainless steel because of exposure to salty air at the base on the Pacific shore, so they did not have the strength of carbon steel cables, but there was more to this problem than that.

It mucho saddened me to know that these cables had never been inspected from the time they were installed, no doubt many years ago. And the installer did not know how to use a tensiometer for tightening. Either that or he or she wouldn't bother to use one.

Those cables had been over-tightened. Just a quick glance at the pulleys told me that. They had flat spots with cable strand indentations. Disgusting. The goofball installer, and every periodic inspector since, had probably never consulted a table of limits. Capital punishment would be too good for them.

At the cable break, I cut away enough length for the splice, and had one side spliced when I was stunned to hear Charlie bark out an oath and start the plane's engine. Why was he doing that? He wasn't a pilot. We had no elevators control, and this small lake required a reverse turn takeoff run to avoid crashing into the trees at the end. I'll explain how that's done in a bit.

And then, to my horror, I decided that for some insane reason he was probably intent on suicide. We would crash into the trees at the end of the lake. He would die instantly and humanely. I would fracture several bones and lay there in extreme agony for hours, maybe even days, until a bear, smelling the blood, would break in and devour me a piece at a time. Life wasn't sweet anymore.

I barked out a protest, but the engine noise was too loud. Charlie also barked out something I couldn't understand. The plane's acceleration, especially as the nose rose to point upward before the floats were up level on the step, caused me to slide backward, wedging my head into the back.

By the time the Cessna was up level on the step I had completed flashing my entire life before me and, except for the image of being eaten, almost achieved the peace of total acceptance.

But then, to my euphoric relief, he pulled back on the throttle. The plane came to a stop as I began crawling backward.

"Grizzly," shouted Charlie.

So okay, I told myself, a grizzly bear was after us. That's why Charlie did it. But I still retained the image of Grizzly Big Fangs eating me piecemeal if he could get to us.

I had a big decision to make. The splice was only half done. Do I interrupt the splicing and crawl out to assess the situation, or do I leave Charlie to cope his best and me finish the splice quickly?

A problem leaving Charlie to cope is that, for one, he could destroy the engine. Another quick start and acceleration might permanently damage the engine, that is, if it had not already been permanently damaged by the first quick start and acceleration.

Under ordinary circumstances, the prop should have been rotated by hand before the start. That would have provided the small spurt of oil to the engine's innards to lessen start-up wear.

A quick acceleration means a steep engine temperature rise before the oil pump has had enough time to not only provide enough lubrication for the moving parts, but also not enough quantity flowing to carry off the excess heat. That would stress all parts, especially the cylinders. That's a good way to crack them.

Another problem with the quick acceleration is that the crankshaft dynamic balancers would deform, leading to excess vibration, higher noise levels, and premature engine failure. Maybe the bear had gone away.

"Mister Grizzly is swimming toward us," squeaked Charlie.

I had to think quickly. Should I take a chance on Charlie burning out the engine, or stay to finish the splice so we could get out of there quicker if we're not eaten first.

Choosing to stay and finish the splice, I told Charlie to get a paddle, go out onto the float, and take evasive action. If he couldn't evade the bear that way, I would quickly climb up front and properly use the engine to evade him, although that could go on all day, leaving us insufficient fuel to get back to base.

Charlie stepped out to paddle and I continued with the splice. About half way done on the second part of the splice, I detected that the plane stopped.

A moment later, Charlie opened the door. "Bear's gone," he intoned between rapid breaths. He had obviously stressed his out of shape, pudgy body. I was hoping he could avoid a heart attack. He climbed into the passenger seat, I finished the splice, and then crawled back to the cockpit.

*

I had just sat in the pilot's seat when we were shocked to see the gigantic grizzly suddenly loom very large, charging out of the trees to our right.

The wiley bear had tried to trick us into thinking he had gone away in the opposite direction, but he had actually looped around for this surprise attack.

As I quickly and aggressively started the engine, the grizzly splashed into the water. He tried to get up onto the right float. The Cessna's nose rose up and the wing tilted over at an awkward angle until the tip was almost in the water.

My imagination raced. I pictured us being found drowned in the submerged cabin of the upside down plane, although I guessed that would be better than being eaten.

Our lives trumped the rapid engine wear, so I slammed the throttle forward. Oddly, I felt an almost perverted pleasure at inflicting damage to the engine.

The plane rocketed forward in a twisted way. Thankfully, the bear tumbled off the float with various banging and fell back into the water.

The plane settled back onto both floats, so I continued the full power. I had to get the plane out of the lake before the bear could get into position to cut off a successful takeoff.

Ordinarily, I would lift the float's water rudders for a takeoff so that the plane could wind cock to face directly into the wind for the most efficient takeoff.

In this case, I needed to do a reverse run takeoff, so I left the water rudders down to first get the floats up on the step and then skid the plane around the lake's far end for a one hundred eighty degree turn to finally head into the wind at speed.

If I had simply started from idle at that far end, there would not have been enough room to get off. We would have slammed into the trees for sure.

However, the problem now was that this second phase of the takeoff would be back toward the grizzly. Charlie held on as we skidded around. After the turn, I saw the bear had manoeuvered himself to be directly in the middle of my path, exactly where I wanted to go.

As an attempt to deke him as in an old football move, I pointed the plane's nose to the left. With dismay, I noticed that the bear did not take the bait. He remained rock solid in the middle, like an excellent linebacker.

So I pointed the nose to that fairly narrow opening to the right and committed to it. We would either make it or become chump for grizzly and his relatives.

Bear sensed my commitment. He began swimming with all his might to cut us off. My takeoff run was lengthened a bit because that deviation to the right was into more of a crosswind.

At the last moment, I yanked back on the yoke. The bear lunged. I swear he was grinning. This was fun for him. As the floats left the water, there came a bang. The Cessna yawed. The trees loomed large. Bear must have swatted the float.

Or had he hit the tail and damaged it? Would I mercifully be unconscious at the first bite – or at least the second?

The floats sliced through the tree tops. The plane slowed a bit, but we remained level just above the tree tops until the increasing airspeed allowed us to gain altitude.

Charlie and I exhaled loudly at the same time. The after effect of the sizeable adrenaline dump into our bloodstream transformed the moment into an exquisite form of nirvana.

*

Back at the seaplane base on the coast, I located the maintenance log and printed in big red letters the cable splice thing. I also found the name of the inspector who did the last annual inspection, found him, and directed unkind words to him.

I should mention that it takes an enormous amount of effort and life consuming time to maintain an IA (Inspection Authorization), that only the most dedicated of people achieve and maintain that precious authorization. This case was extremely rare. I don't know what had happened to him to allow such negligent, dangerous sloppiness.

It 'bears' mentioning that the whole fleet was grounded until the proper inspections and repairs were made."

Please note that most of the students in the class said they did not believe the story above, but found it entertaining and instructive. I wonder how many of them henceforth included a small length of cable in their tool box. I did.

Chapter 25

On To the End

The weeks and months after that first semester carried on less dramatically while we learned more wonderful, technical things from the old pros. Most of us attended classes right through the hot Florida summer so we could either graduate early and start earning an income in the real world, or carry on to engineering.

We learned that some aviation businesses operating with meager funds might want us to take shortcuts and sign off work and airplanes that are not actually safe for flying, that we could be charged with criminal negligence, or that manslaughter or worse with the appropriate prison time could happen if we give in to it. Such knowledge makes us less afraid of being fired for insisting on safety.

Later, in the 1970s after the world oil crisis when fuel costs rocketed up, the air regulations relaxed somewhat as a concession to the general aviation business, so that aircraft owners could save money and keep flying by performing maintenance on their own airplanes.

The work was/is should be performed under the supervision of a trained mechanic, and an IA (mechanic with an inspection authorization) still had to sign off the annual.

For the annual, the inspector is supposed to be right there taking part in the inspection step by step, but sometimes the rule is not applied evenly. I'm referring here to the general aviation of small planes. The airlines don't have annual inspections. They have progressive inspections throughout the year. All amateurs must bug out. Only pros allowed. That's why airliners are so relatively safe now.

It's then that I wonder about how maintained those smaller, general aviation planes are. There are certain practices that technicians are trained in, but that owner/pilots may not know.

An example is the tightening of a connection such as with fuel or hydraulic lines. Depressing things can happen if such line connections leak.

Each and every time such lines are connected, a calibrated torque wrench is to be used. Calibration means that the wrench is officially tested for accuracy at regular intervals at an officially approved facility.

Some people, even licensed techs, think they can simply 'feel' the correct tightness without using a torque wrench, so they make the connection, pressure test it, and it's okay.

What usually happens with 'feel' tightening is that the connection is either too tight or too loose. What may work in other industrial applications will fail on an airplane because of the relative vibration and large, ambient temperature changes.

If a connection is too loose, vibration and temp change will loosen it after a while and leak.

If it's too tight, the connection will not absorb vibrations. The stress concentration cycles will work harden the connector, leading to cracks, leading to failure or a leak.

The correct torque for a particular connection is the 'sweet spot', in which the connection will not loosen and the vibrations and heat expansion/contraction will be safely absorbed. It's too subtle for anyone to feel. There are other such applications.

An inspector, if not doing the torqueing him or herself, should at least observe it being done. If torqueing is not observed, the inspector can't simply apply a torque wrench to check the tightness because there's a difference between the static and dynamic coefficients of torqueing friction. A simple torque check would be inaccurate.

*

Another example is if an owner/pilot services the engine's spark plugs. An experienced technician will know that a spark plug, after cleaning and gapping, is moved to the next cylinder in the firing order only if the engine has an even number of cylinders as with horizontally opposed engines, but the plugs will stay in their original cylinder if there's an uneven number of cylinders such as on radial engines.

The reason that happens is that the magnetos fire opposite polarity consecutively – electrode to ground, and then ground to electrode. A small amount of metal is transferred each time. The plug will obviously fail twice as early if that metal transfer is in only one direction. That failure usually occurs when you're flying over a swampy area with lots and lots of predators, especially grizzlies, who become very excited at the sound of a sputtering engine.

Would an owner know to inspect a plug's firing end for color of the deposits? It would reveal to the owner that the airplane was operating too rich or too lean or burning oil.

We were taught that an inspection is a tactile experience, not just observation. We push and pull on things as we go along to check for looseness or unusual noises, and so on. We were taught to see the aircraft in its dynamic environment of temperature changes, vibration, corrosion and other deterioration forces over time. It's with that view that aircraft are to be preventatively maintained, not just fixed.

On the job training would be wonderful, but that's not how most aviation schools are set up.

We were instructed that the college, along with the many pro insights, was an excellent start. Some would say it should be the only start because, with an on the job start, you might get only the insights of one person, and a limited exposure to types of aircraft and systems.

We must expect that we would be learning for the rest of our careers, because, for one thing, the technology is constantly changing and we often would be working from first principles. Boredom need not be in the vocabulary of a licensed aircraft tech.

Our final subject was weight and balance. Then we were to perform a practical project and be verbally tested before going on to the FAA office in Miami to write the series of tests for our license.

Naturally, Fadi was the only one of our group who had a problem. We were to fit and weld together a cluster of chromoly tubing.

Fadi did not like the appearance of his weld, so he applied the torch to it after the initial welds had cooled. Fadi should have known that going back over a weld like that causes carbon embrittlement leading to certain premature failure after vibration and other stresses have their way with it. Anybody, and even their dog, could spot carbon embrittlement.

I don't know why Fadi would do such a dumb thing. It was as though he had a hole in his otherwise very intelligent brain, not a good attribute for an aircraft techy.

The one area in which he did not have a hole was in his ability to perceive he was in trouble and talk his way out of it.

Seeing he was about to fail the practical part of his examination, he, with a wide smile, explained to the examiner that he would not be working on airplanes, that he only took the course because he loved airplanes, but he would be working for his father's merchant business back home in another land.

The examiner blinked a couple of times, sighed and generously gave him the chance to weld another cluster. So he did that and passed.

The exams went well for me at the Miami FAA office. I received my licence. At our Embry Riddle graduation, it was stressed that although our license said 'mechanic', we were actually technicians and that term was on our diplomas. A mechanic turns a wrench, but a technician figures out from first principles why a problem developed and how to prevent future problems.

Our group said our goodbyes, promised to stay in touch for the rest of our lives, and set off for home in anticipation of the real world of aviation.

Epilogue

Every December I fly on my flight simulator program an Eastern Airlines Super Constellation from Miami, up the eastern coast to Philadelphia, and on to LaGuardia in NYC where I fuel up for the completion of the flight to Chicago that I had experienced after the first semester all those decades ago. From there, I cheat and fly a seaplane to the Sault.

I fly a seaplane, a Grumman Goose (with growling twin R985 radial engines}, because the Sault did not have an international airport like it does now. I land on the Saint Mary's River at the location where the Algoma Flying Club used to be.

Just before taking off from Miami International runway 27, I glance out the left windshield to catch a glimpse of those four young, Embry Riddle guys parked in a Caddy with the top down, fully engaged with all their emotional being in the excitement of the aviation activity they were observing.

After takeoff, I turn to fly low over the Embry Riddle campus as the propliners did so spectacularly in the young man's sight out the dorm window that first evening in Miami. I pretend it had not become a juvenile detention facility.

I remember crawling under the college's back fence, meeting Nanny, seeing the boot salesman instructor going bonkers, Tiny punching Fadi, seeing Fadi dangerously pretending to be a gunslinger with the rivet gun, great insights from the instructors, and a flood of other such memories.

I spot the hotel on Flagler. I remember the crashes against the wall and hearing "Yip, yip, yip", walking with a severe slant into the wind of a hurricane, Fadi barking directions to Tiny in the Caddy and huge Bartholomew's tentative request for Tiny and I to be his and Fadi's roomies.

I tip a wing to see Barry College. Bartholomew dances with a 'Nun'. Tiny, unaware of his love's unfaithfulness, is faithful. Fadi dances with the Queen Bee, or with whom he thought to be the Queen Bee.

Personally, I wonder if the veteran jet fighter pilot, Charlie, had only been joking about there being a Queen Bee guarded by a Nun. It may have been only chance that two students looked the part at the dance. Maybe Charlie had not even been a jet fighter pilot. Hmmm.

I pass by the airport to bid a goodbye to the Silver Springs complex. It's gone now. I picture Eddie Rickenbacker across the street from the complex pointing a crooked finger at the Eastern's board, warning them that jetliners would ruin the company. I see a line of L-188 Electras at the maintenance bays.

From there I carry on up the coast to Philadelphia, then toward NYC's LaGuardia. I fly the V123 using ADF as my secondary reference. I try to not think of the terrible crash of the two airliners that occurred on that date.

After fueling at LaGuardia, I take off for the leg to Chicago's O'Hare. From there, instead of taking the bus, I switch to a seaplane, a Grumman Goose with two growling R985 radial engines. I make that departure from the actual memory to speed up the trip and land on the St. Mary's River at the Sault.

Before heading north from O'Hare in the Goose, I spot the Chicago bus terminal where the larcenous taxi driver parked in front of the mounted policeman.

Then I carry on to the Sault, finally landing on the river where the old Algoma Flying Club used to be. The property now has an enormous mansion on what is now the expanded Anishinaabe reservation.

*

Tiny married the Barry College student he dated after the breakup with his former love. They had children and have remained happily married. His former love did not go on to have a happy life, but Tiny has never gloated.

He found employment with a large aviation company where he rose into middle management. He liked the job, but had to volunteer his occasional maintenance services at a local FBO to maintain his psychological balance.

Bartholomew and Rosemary remained devoted to each other. The apparent age difference has not been a problem. For a while at the beginning, they made use of Bartholomew's aviation training in their ecumenical church directed aid work overseas.

They then returned home to the States where Bartholomew became a pastor, and Rosemary later became a pastor as well, although she maintained a teaching position.

It became an ongoing joke with us about whether or not Rosemary had been what we took to be a Nun at the Barry College dance. I still don't know for sure, but I think so, but maybe not.

Fadi married that Barry College student from their home country, raised a family and stayed devoted to each other. He did work at the family business, eventually taking it over when his father retired.

Like Tiny, Fadi had to indulge his aviation passion at a local airport, except he was well enough off to buy his own plane, which almost killed him.

He insisted on rigging the controls himself, but rigged the ailerons backward. Right after takeoff, he tried to turn to port, but his brain couldn't handle the reversal. He cartwheeled into an orchard.

As with his cluster weld fiasco at the college practical exam, and the gunslinger episode, he came to realize he had as if a certain hole in his otherwise very intelligent brain where only a vacuum existed.

His family, especially his new wife, forbade him to ever again perform maintenance on his or anyone else's plane. Apparently, his flying remained exempt from that logic hole in his head. I'm so glad he never put others' lives at risk.

I was the only one to not marry a former Barry College student (or Nun). I married a wonderful young woman back in

Canada for a happy, lifetime commitment. The guys want to remain relatively anonymous.

Embry Riddle went on to become a stand-alone university with several campuses, the largest ones in Daytona Beach, Florida and Arizona. It still retains the world renowned reputation of aviation excellence that it had those decades ago in Miami.

The real world of aviation now beckoned to me.

The author back then. He doesn't look like that anymore.

Embry Riddle, Airframe and Powerplant Technology, A&P licence. York Univ, BAS. Univ of Toronto, MAp

FOB and Douglas structural experience. A course developer and trainer for Douglas DC-10 program. Developer and instructor for de Havilland Aircraft Dash-7 systems. Developer and instructor for Centennial College aviation program.

www.ingramcontent.com/pod-product-compliance
Lightning Source LLC
Chambersburg PA
CBHW071428180526
45170CB00001B/264